ISBN: 0615778194
ISBN-13: 9780615778198

Library of Congress Control Number: 2013934812
Franklin Square Publishing
Washington DISTRICT OF COLUMBIA

In memory of Edwin R. "Tony" Anthony, 1948–2011.

He cared deeply about the grid electricity industry and inspired very many of us.

Lots of pow'r to spare

It's heresy to bring up Reddy Kilowatt these days. The cute mascot, popular when America built most of today's grid infrastructure in the half-century of 1940–1990, has been persona non grata in the twenty-plus years since.[1]

Why isn't Reddy politically correct? What does his banishment some twenty years ago say about the grid's direction? Have we lost our way without the little man to guide us?

He's cute in a 1950s way, harkening back to the heyday of "Father Knows Best," Ike, Mr. Machine, and hula hoops. His wiry body, as trademarked by A.B. Collins in 1935, is composed of cartoon lightning bolts. His warm face has electric outlets for ears and a light bulb for a nose.

His 1959 song was silly but suitable for those times:

> I'm a real live wire, and I never tire. Yes, sir!
> I can cook your meals, turn the fact'ry wheels,
> 'Cause I'm RED-DY KIL-O-WATT.
> When you toast your toast or you roast your roast, it is I who makes 'em hot.
> I'm so full of spark, I can light up the dark.
> I wash and dry your clothes, play your radios.
> I can heat your coffee pot, I am always there—with lots of pow'r to spare.
> 'Cause I'm RED-DY KIL-O-WATT.

With Reddy urging us on, the country wired itself up in the middle of the twentieth century. Yet the spirit of the grid guy is nearly forgotten now. Nobody talks about him anymore. Nobody *wants* to talk about him anymore.

---

1 The Smithsonian National Museum of American History holds thirty cubic feet of Reddy Kilowatt advertising materials, business correspondence, newspaper clippings, ephemera, radio scripts, speeches, film, videos, audio cassette tapes, audio discs, filmstrips, compact discs, photographs, and scrapbooks at the museum's Archives Center, in 119 boxes, covering the period of 1926–1999.

Even mentioning Reddy draws cross stares. Whispering his name suggests you condone the promotion of electricity when you know you must condemn it.

For, some twenty years ago, it was decided that grid electricity should never, ever be promoted. It was decided that America should stop growing the grid.

Light your house, charge up your devices, watch your favorite shows. Manufacture the product, cool the store. But do it with as little electricity as possible.

Utilities took on the job of telling their customers to ration kilowatt-hours as if in wartime. This was crushing news for Reddy fans.

There was some experimenting, to have Reddy act as a cheerleader for rationing. It didn't work, and those ideas have since been abandoned. Reddy simply couldn't do it. His first and always true love was grid electricity.

It's painful when the lights go out for a few minutes, more so when they stay out for a few hours. But we count on the electricity system righting itself before much longer, no matter how hard nature hits us, no matter what it takes.

"After the electric light goes into general use
none but the extravagant will burn tallow candles."

Thomas Alva Edison,
quoted by the *New York Herald*, January 4, 1880

Contents

List Of Graphs

Acknowledgments

The inspiration for this book came to me as a gift from friends and colleagues in the electricity profession. Some of the smartest thinkers of my generation, about energy, couldn't have realized they were giving me something precious. But they did.

They patiently worked through with me this new "value of electricity" perspective. Value of electricity, that's what people are now calling it. We prodded and tested it in long conversations. Would these new principles withstand the scrutiny sure to come? And we talked out whether their adoption and application would have the good effect that I intended.

Already, value principles are being incorporated into industry strategies and regulatory policies. This is satisfying to be sure, hopefully for those who brainstormed with me as well.

However, the book is entirely mine. I alone am responsible. Particularly because the book hurls a few lances of logic at fashionable assumptions and conventions.

First, I need to acknowledge the contribution of my friends and colleagues at Bates White Economic Consulting, where I am a part-time senior energy advisor. Bates White has been extraordinarily generous, providing me virtually unlimited research resources (economists, engineers, statisticians, computer resources for data analysis, clerical support, etc.).

Dr. David DeRamus, Dr. Kathleen King, and Nick Puga, leading consultants at Bates White, constantly spurred me toward rewarding lines of thinking and research. This book would not have been possible without their brilliant insights. Others there that provided critical ideas and help include Natalie Fike, Vicki Marone, Austen Talbot, and Dr. Spencer Yang.

Next, I need to acknowledge the contribution of my friends and colleagues at the Edison Electric Institute. The EEI must walk a fine line balancing competing interests among member utilities, which it does exquisitely and sensitively. The EEI and member utilities will not embrace many of the premises and assertions herein. That being said, the thoughtful inputs of these leading thinkers were essential to where the book comes out. No matter your agenda, you have to see that the EEI plays one of the most important and noblest roles of any business association on behalf of all Americans.

Then, I need to acknowledge the contribution of many friends and colleagues who lead or have led America's electric utilities. Leading an electric utility in the late twentieth century and early twenty-first century has not been an easy task by any means. Our society puts enormous pressure on these people to satisfy many conflicting expectations. It's literally impossible to meet all of the expectations amid extreme resource constraints. Utility leaders are rarely celebrated and often berated.

If the EEI must walk a fine line in what it says and does, electric utility leaders must walk a barely visible line. One wrong word can set off a public controversy and cost the utility big-time. The popular impression is that utilities are monopolies that can say and do anything they want. This couldn't be further from the truth.

Half business and half dictate of the state, utilities are entirely dependent upon the good will and good sense of their regulators. The regulators are, in turn, entirely dependent upon the support of political leaders and ultimately the public. For those who think that utilities are immune to the needs and wants of its customers, unlike unregulated businesses, think again. Better yet, read this book.

For this reason, I'm not able to acknowledge by name the utility leaders who made a difference in the evolution of the value of electricity perspective. But I do feel obliged to acknowledge their decisive intellectual input.

Finally, I need to acknowledge leaders in utility regulation who made a difference in the formation of the value of electricity perspective. They would not agree with all of the conclusions that resulted from my research. But they probably would agree on their usefulness in regulatory debates.

After thirty-five years in the electricity profession, I recall many others along the way who aided my career and helped put me in a unique position to formulate and communicate to a wider audience the value of electricity. It's a great pleasure for me to acknowledge these friends, colleagues, assistants, mentors, bosses, and clients.

The electricity profession is a peculiar field, highly regulated and politicized in almost every regard; arguments involving slews of lawyers and technical experts and thousands of pages of "documentation" erupt at every turn.

It has often seemed to me that the number of us white-collar pencil pushers is way out of proportion to the number of engineers and linemen out there. It's them, not us, who track fearsome storms, reassembling afterwards the circuits that power our computers, cell phones, and air conditioners. So I would be remiss if I did not close the acknowledgments in this book by honoring the great Americans in the International Brotherhood of Electrical Workers (IBEW), in other unions, and those not in unions who extend and defend the grid. These good people inspire me each day.

Author's Note

This book demonstrates how averages can be poor measures of how Americans pay, use, and value grid electricity. Nonetheless, you'll see averages within. Why?

Some sections of the book talk about my first tentative steps into the land of electricity-nomics before the critical limitations of averages became apparent. Be patient. Subsequent sections use more representative statistics and highlight how medians, modes, tails, quintiles, etc. help us better understand what's up with grid electricity.

You'll also see averages late in the book because of a lack of detailed data on some topics. The superior statistics require detailed data, unlike averages. My apologies in advance.

Speaking of averages, the book sometimes talks about the prices of grid electricity and its rates interchangeably. This might disturb an energy expert who frets about the many nuances of rates. For that reader, please bear with me. This book is intentionally not technical. When it refers to prices and rates, the book generally means average prices and average rates, including all of their component parts.

Uniquely, my research team has deeply mined the detailed data that is publicly available as well as much that is not. You'll see hypothetical examples, however, when you'll be hungry for some data-backed analysis. Sorry, but even we didn't have access to detailed data on some topics.

You might also be hungry for more answers. How can utilities, specifically, make the power outages caused by severe storms rarer and less lengthy? How can utility regulators, again specifically, evaluate utility proposals to do so? What is the value to the public of something closer to "perfect power"? What are the preferred regulatory mechanisms to make the costs of more perfect power affordable?

These are all important questions. But the answers fall outside the scope of this book. Instead, *Lines Down* urges us to get down to work. Utilities and utility regulators, both of whom I admire, can and should do this. Let's take the already high value of grid electricity and boost it even further.

The universal access to as much nearly free grid electricity as is needed and wanted empowers America. It enables Americans to care, create, collaborate, and craft with unprecedented efficiency and productivity.

Preface: There's Electricity and Then There's *Grid* Electricity

Grid electricity's nothingness

The Age of Electricity is 130 years old and counting. Every decade brings along with it new machines, appliances, and devices animated by electricity. Electricity's killer apps of yesteryear, like televisions and refrigerators, become bigger and better at satisfying our needs and wants. Today's killer apps, like smart phones and the Internet, become essential sensations overnight.[2]

What feature makes grid electricity so valuable to us? The answer is nothing. That's right. Grid electricity's essential feature is its nothingness.

Perhaps Thomas Edison's biggest breakthrough was delivering electricity to households and businesses in a manner that customers don't sense. Edison's customers knew electricity mostly by the work it performed for them.

What is electricity? Where is it made? How did it get to my building? Edison's customers generally did not know and did not care.

But Edison wanted it that way. He wanted customers to give scant thought to the behind-the-scenes of electricity. He wanted them instead to enjoy the play.

This is a big point. Edison didn't have to go in that direction. To give the superpower of lighting to his Lower Manhattan customers, Edison didn't have to assemble a "grid." But he did.[3]

---

2 Stephen Baker, "Can Social Media Sell Soap?" *New York Times*, Jan. 6, 2013. "Dave Morgan, a pioneer in Internet advertising… points to the early years of electricity. In the late 19th century, most people associated the new industry with one extremely valuable service: light… What these people missed was that electricity, far beyond light, was a platform for a host of new industries. Over the following years, entrepreneurs would come up with appliances—today we might call them 'apps'—for vacuuming, laundry and eventually radio and television. Huge industries grew on the electricity platform. If you think of Apple in this context, it's a $496 billion company that builds the latest generation of electricity apps."

3 Ernest Freeberg, *The Age of Edison*, Penguin Press, 2013. "[Edison] came to envision not just a new type of incandescent lamp, but a central station capable of distributing light and power over many city blocks… His improved dynamos would send current through buried copper wire mains and into each house the same way gas traveled… Both Edison and his rival electricians sold stand-alone systems, single dynamos that fired a string of lamps, enough for a large house, store, or ship. But after the successful test of his Pearl Street station, Edison hoped to move forward with his much grander vision for an electrical grid, installing his central station in the urban core of every major city."

Edison decided to manufacture electricity off site. He figured customers would embrace electricity if all of the grime, noise, and nuisance were out of sight, out of mind. He was right. That's how electricity would be most valuable for people.[4]

Cannot see, hear, smell, taste, touch it

Electricity is so valued because it's the shyest of energy sources when it comes from the grid. So shy, such a wallflower, that it's practically nada.

You cannot see grid electricity. You cannot hear or smell or taste or touch it. It is there, but it's not. Electricity is performing observable work for you: lighting the room, bringing your computer to life, cooling your milk, etc. But aside from this impressive output, it isn't here or there.

Visualize Isaac Newton, James Watt, or Benjamin Franklin time-traveling to our present-day world. After arriving, they acclimate to our modern bustle and brutishness in remarkably short order.

At first they puzzle about our strange dress, music, speech patterns, and affinity for holding talking machines in our hands. We ask about their reaction to electricity. Our giants of scientific thought are most fascinated by how difficult it is to detect electricity's presence. They might ask us, "How do you know you have some electricity?"

These great men routinely observed and studied energy performing work for themselves and the people around them. However, in their own time, they never once witnessed energy performing work that, like grid electricity, hardly leaves a ghostly trace.

They never imagined energy performing work without flame, smoke, noise, and hulking machinery. They never imagined energy performing work without anyone, right there, tending to the energy-making where the work is consumed. Yes, no doubt that Newton, Watt, and Franklin would find electricity's nothingness fascinating.

4 Freeberg (2013). "Although Edison's incandescent lighting system was one of the most sophisticated pieces of technology yet created, all of the complexity had been engineered out of sight, invisible to the consumer... Oil lamps and candles required wick trimming and soot cleaning, while gas burners demanded even more technical skill from consumers, who had to adjust meters and burners in addition to regular cleaning. But the electric light required no maintenance, while the source of the power hummed out of sight, sometimes many city blocks away."

Scavenger hunt

I exaggerate slightly about the nothingness of grid electricity. Search around your home. You can seek out and spot clues that lead back to electricity. How many telltale switches, wall plugs, insulated wires, and chargers do you find dispersed about? Then enlarge the manhunt. You come across the fuse box in your basement and utility meter in your backyard.

Look above, now that you're outside. You trace with your eyes the overhead power wire snaking from the street, through the trees, and downward to the utility meter.

That power wire doesn't seem like much, does it? It's not much more than the wire that plugs the toaster into a kitchen outlet to brown your bagel. But that one wire is an umbilical cord to the nation's electricity system. It brings to your family virtually unlimited energy, courtesy of our grid. Your family can use that energy, however much they want, however they see fit.

Take ten kilowatt-hours from our grid today. Take two hundred kilowatt-hours tomorrow. That's twenty times as much, no questions asked. That quiet wire up there is at your command. It's a big open pipe for as much energy as you order up.

The central air conditioner cycles, automatically, to maintain a constant interior temperature regardless of the outside temperature. The washer and dryer start up instantaneously at your command by lightly depressing a button with a fingertip. Devices throughout the home snap to attention and operate as their master, you, wills it. And portable electronic devices receive their lifeblood, all without a flame, a windmill, or a beast of burden in your living room.

The universal access to as much nearly free grid electricity as is needed and wanted empowers America. It enables Americans to care, create, collaborate, and craft with unprecedented efficiency and productivity.

We're increasingly able to concentrate on activities we prize in comfortable spaces anywhere. More and more of us are safe, healthy, and functional for nearly a century of life. We're increasingly mind melding, via the Internet, with whomever we want wherever they happen to be. We're increasingly delegating drudgery and distraction to machines, appliances, and devices.

Don't know what you've got 'til it's gone

Then suddenly it's gone, all of it at once. How appropriate that we call this a blackout, as if our everyday lives go black, as if we are knocked into unconsciousness by a bloodied prizefighter.

I was a thirteen-year-old in Brooklyn when the famous Northeast Blackout of 1965 struck. Someone set a relay too low near the Niagara hydroelectric plant in Ontario. At 5:27 p.m. on November 9, 1965, some 30 million people were reminded of the magic of electricity and our grid. They lost it, for what seemed to be an eternity at the time, as long as twelve hours for some households and businesses.

WABC's Dan Ingram was doing his afternoon drive-time radio show. At 5:21 p.m., he noticed that the music he was playing was slowing down. At 5:25 p.m., Dan told listeners that the electricity seemed to be slowing down. He didn't know it could. It was "slowing" from 60 down to 56 Hertz six minutes before the blackout and then to 51 Hertz two minutes beforehand.

For a thirteen-year-old in Brooklyn, the Northeast Blackout was great fun: doing my homework by candlelight, the family huddled together with nervous excitement, made a little bit cozier by the quieted living-room TV, the rabbit ears antenna rendered powerless. Hailing the grid's return, the lights turned themselves back on, after the five-and-a-half-hour intermission, at a few minutes after 11:00 p.m.

The Northeast Blackout was famous in its day. It was unprecedented, both in terms of the number of people affected and how long it lasted. People talked about the blackout for weeks and months afterwards.

Yet the average duration was around six hours. And everyone's service was restored within twelve hours or so. Can you believe that? Today, an outage of that relatively short duration would hardly be a blip on the screen.

Multi-day pain

Nearly fifty years later, a blackout has become a more serious matter. A blackout today is a catastrophic disruption felt in every workplace, every home.[5] In fact, a blackout is such

5 At the time of the 1965 Northeast Blackout, 39 million American households connected to the grid took about thirteen kilowatt-hours daily, on average. Today, for the 115 million households connected, the average is thirty-one kilowatt-hours daily. Source: US Census Bureau, Statistical Abstract.

a serious matter that an intentional one is a top national security threat. A cyber attack on the grid could black us out, taking American lives. The occasional blackouts were once accepted as a fact of life. Nothing you could do about them, cost-effectively.

Now, multi-day blackouts are failures we cannot afford.[6] Medical devices shut down. Phones drain their batteries. ATM and checkout machines freeze. Cooling and heating systems go quiet in our airtight, overly insulated buildings. Our Internet-based society goes on a wildcat strike.

We count on the grid working without interruption. It's painful when the lights go out for even a few minutes, more so when they stay out for a few hours. But we count on the electricity system righting itself before much longer, no matter how hard nature hits us, no matter what it takes.

Count on me

The grid, in turn, counts on us. It needs us to hear it, though it's silent. It needs us to see it, though it's invisible. It needs us to remember it, though it's complicated and elusive.

It needs us to devote sufficient resources. For if we scrimp, the grid will increasingly let us down. If we allocate more resources to fashionable initiatives, leaving less for the grid's basic mission, its unique qualities will inevitably degrade.

Those on the "soft path" bandwagon are remarkably unconcerned. A few even welcome the grid's upending. They're enthused about the reallocation of manpower, management, and money toward the fashionable and away from the basics.[7]

---

6 Office of Maryland Governor Martin O'Malley, "Weathering the Storm, Report of the Grid Resiliency Task Force," Sept. 24, 2012. "Power outages represent significant inconvenience on many fronts, including interruption of Internet access, possible interruption of telephone service... Perhaps of greatest importance is the health-related impacts to the elderly and infirm who are particularly susceptible to extreme heat and cold. Compounding the potential adverse health consequences is the chance that emergency services provided by local and State government also may be compromised during periods of outage."

7 Amory B. Lovins, "Energy Strategy: The Road Not Taken?" *Foreign Affairs*, October 1976. This oft-cited article by grid electricity's most famous and effective critic includes some rather radical thoughts: "For some applications, electricity is appropriate and indispensible: electronics, smelting, subways, most lighting, some kinds of mechanical work, and a few more. But these uses are already oversupplied, and for the other, dominant uses remaining in our energy economy this special form of energy cannot give us our money's worth... Thus an affluent industrial economy could advantageously operate with *no* central power stations at all! ... Civilization in this country, according to some, would be inconceivable if we used only, say, *half* as much electricity as now. But this is what we did use in 1963, when we were at least half as civilized as now. What would life be like at the per capita levels of primary energy that we had *in 1910*...but with doubled efficiency and with important but not very energy-intensive amenities we lacked in 1910, such as telecommunications and modern medicine?" *Emphasis added*

This book is, ultimately, deeply concerned about maintaining grid electricity's vital qualities and value. But this book is also optimistic that Americans will rediscover the grid's greatness. With that rediscovery will come, I think, the need to revitalize it, to shore up grid quality, to maintain its value.

Steven Allen Mitnick
Washington, DC
June 2013

The Northeast Blackout of 1965 was famous in its day. Yet, the average duration was around six hours. Today, an outage of that relatively short duration would be hardly a blip on the screen.

Introduction: Electricity Is Good, Kinda

Electricity is good!

I designed and ordered the buttons on a web site. They had a bright white background, seventeen stark blue letters, and an exclamation point. A hundred buttons, each with the unexpected message: "Electricity is good!"

It was the spring of 2011. The electric utility industry's annual convention was coming up. I was armed and loaded...with buttons.

Hungrily taken by the industry's leaders, fastened to many lapels, the buttons nonetheless drew the same quizzical look every time. What's up with this message? What are you trying to say here?

Reaction to the button went like this. Sure, electricity is good. Well, kinda. But to come right out and say it, that's pretty edgy. The industry's leaders could never do that.

Saluting our grid by wearing a simple button was a bit retro, a bit naughty. Harkening to a time long past when electricity was a product to be proud about incontestably, not a necessary societal evil. When delivering electricity amply, economically, safely, and continuously, via our grid, was a service to be equally proud about.[8]

There was a time not long ago when our grid was an American marvel. When America knew our grid needed to be well cared for and grow ceaselessly to advance our freedom, our strength, our spirit, and our way of life.

---

8 Juliet Eilperin, "Study: Proliferation of Home Air Conditioners Saves Lives," *Washington Post*, Dec. 23, 2012. "The installation of air conditioning in American homes is the reason why the chances of dying on an extremely hot day fell 80 percent over the past half-century, according to an analysis by a team of American researchers... A team of researchers from Tulane University, Carnegie Mellon University, the National Bureau of Economic Research and the Massachusetts Institute of Technology examined patterns in heat-related deaths between 1900 and 2004. The group found that days on which temperatures rose above 90 degrees Fahrenheit accounted for about 600 premature deaths annually between 1960 and 2004, one-sixth as many as would have occurred under pre-1960 conditions."

Personal journey

Able to follow my passion in early 2011, I founded Build Energy America to do good research and advocate for investment in the nation's energy infrastructure. Many friends knew of my passion. I had founded a company following 9/11 to upgrade the transmission of electricity into New York City. I had joined Eliot Spitzer when he became New York's governor, with his commitment that my job as "energy czar" would be to get energy infrastructure built.

Out of the gate, Build Energy America documented the enormous potential for job creation from increased energy infrastructure investment. The report attracted interest from Democrats and Republicans, both in pursuit of job creation strategies as the presidential election of 2012 approached. It was nice to have an impact on the thinking of political leaders.

Then I hit the wall.

That wall was utility regulation, and it was solid brick. I didn't bounce off so easily.

Friends in utility regulation clued me in. It almost doesn't matter how compelling a proposal is for increased energy investment. Even if the greater investment would bring with it hundreds or thousands of new jobs, it does not matter. You can just forget it. It's not happening.

How could this be? You mean even if the greater investment would fill a real need to modernize aged assets or grow capability? You mean even if the greater investment would yield benefits far in excess of costs? How could this be?

Haven't you noticed, Steve? The country is in a deep recession, they reminded me. People are hurting. Many can't make ends meet. Part of their pain comes from sky-high electricity costs that keep going up and up. So there's no way utility regulators are going to approve industry proposals that would add to the pain.

Talk about a barrier to increased energy investment. I might as well change the name of my research and advocacy organization to Broke Energy America. Or Don't Bother Building Energy America. How about Build Energy China?

Then it happened. My personal journey began that day. It was March 2011. It occurred to me that what my utility regulator friends were advising me was true but not true too, kinda.

They were right that utility regulators were in no mood to approve industry proposals. They were right that regulators' reluctance was due to concerns about electricity costs and the burden on households and businesses.

But my regulator friends were wrong about an essential point. Electricity costs aren't sky-high. I knew they weren't. It made me wonder, why is there such a gap between the facts about electricity costs and the general perception?

The gap remarkably applies to utility regulators as well. Most regulators are extremely knowledgeable about the energy industries. So why do savvy regulators, along with the public, have a conviction that electricity costs are sky-high and going up and up?

Up, up, up

That's when I knew. Build Energy America just had to get to the bottom of these questions, rigorously, convincingly, before tackling anything else.

Are electricity costs sky-high, as is often said? Or are they moderate? Are electricity costs going up and up, as is often said? Or have they been stable?

For if electricity costs are moderate, if they've been stable and are likely to remain so, then why doesn't the public know this? And why don't utility regulators know this? That's really troubling. You would expect regulators, of all people, to be totally on top of electricity costs. How can they be so far astray in their very own field?

If electricity costs have been consistently moderate for decades then why don't regulators factor this in as they set the grid's course? Why do regulators act as if the costs to keep our grid solid are much too painful to bear?

So in June 2011, after researching electricity costs from every direction, Build Energy America published "Electric Values." I subtitled the report, "Ten Quantitative Analyses Challenging the Myth that Electric Bills Are High and 'Going Up, Up, Up.'"

Okay, it was neither succinct nor catchy. But the subtitle had special meaning for me. I had seen the "Going Up, Up, Up" part in a *USA Today* headline. The careless headline screamed, "Price Jolt: Electricity Bills Going Up, Up, Up." For me, the headline represented what we're up against. How could *USA Today* shamelessly run this in its papers?

Skews everywhere

Then everybody in the field stopped what they were doing. In unison they said, "You're right, Steve. How could we have missed this? Electricity costs actually have been consistently moderate for decades, just like you have it in the report. Gosh. We'll make a complete one-eighty, knowing now that grid electricity has been, is, and will forever be a great value for the public."

Uh, that's not really what happened when "Electric Values" came out. No one could find fault with the report's logic and evidence. But life, such as it is in the electricity business, went on as before.

Did I give up? You know I didn't. The stubbornness of the electricity business's myths didn't drive me to drink. It pushed me to more research. I was joined in my journey by an increasing team of talented economists and engineers, in an analytic growth spurt, and we turned up discovery after discovery. Each new finding shook up experts and laymen alike about how much electricity Americans really do take from the grid, how much they pay for what they take, and how much value they get. It's been a wild ride indeed.

The door opened on this new world of "electricitynomics" when, in August 2011, the team made the key finding. A finding so simple, you have to wonder why we or any expert didn't see it before. Like so many other breakthroughs in history, simplicity and complexity joined.

Electricity is skewed. There, I said it. Look at the public's electricity usage. Look at the public's utility bills. Look at how utility blackouts occur or how the public incurs costs from blackouts. There are skews everywhere. How all of these phenomena are statistically distributed is skewed rather than normally shaped.

Once the team had this insight, our research took off like a rocket. What you, the reader, find within these pages are the fruits of, essentially, that discovery in August 2011 of electricity's skewedness.

Saluting our grid by wearing a simple button was a bit retro, a bit naughty. Harkening to a time long past when electricity was a product to be proud about, incontestably, not a necessary societal evil.

Grid Electricity Questions

On why grid electricity is believed to be expensive and becoming more so, though it's not, and how its expense should be measured.

PART I

Part I. Grid Electricity Questions

Chapter 1: Is It a Bad or Good Deal?

Chapter 2: The Value-for-Money Question

What's up?

USA Today's headline was provocative: "Price Jolt: Electricity Bills Going Up, Up, Up." The article was typically dramatic.[9]

Who wouldn't be anxious after seeing this? Nobody would notice—years later—when the article's grim forecast was unrealized.

The US Energy Information Administration reports that households paid 4.1 percent more for a kilowatt-hour of grid electricity in 2011 than in 2008, when the article was published. So, over the course of three years, the cumulative "price jolt" was 4.1 percent.

This number exaggerates the so-called jolt. General inflation is ignored. The Consumer Price Index, a metric of inflation, increased by 4.5 percent during the same three-year period. This means that the 4.1-percent increase was really a slight decrease, appropriately adjusted for inflation. The price jolt wasn't even a spark.

USA Today is hardly alone. The media constantly reports massive rate hikes that later turn out to be a mirage. It constantly pitches that consumers—already heavily burdened by high rates—face ever more rounds of backbreaking rate increases.[10]

---

9 Paul Davidson, "Price Jolt: Electricity Bills Going Up, Up, Up," *USA Today*, June 20, 2008. "Here's a shocker: Electricity bills are heading up. Way up." Yet, according to the Energy Information Administration, American households' average electricity price in 2008 was 11.26 cents per kilowatt-hour and it increased 0.25 cents in 2009, 0.03 cents in 2010, and 0.18 cents in 2011, all without adjusting for inflation. In real inflation-adjusted terms, there was no price jolt.

10 Jonathan Fahey, "Electricity Prices Rise despite Cheaper Costs for Utility Companies," *Huffington Post*, July 11, 2012. Despite the headline, the article admits: "Electricity prices are forecast to rise *slightly* this summer. But *any* increase is noteworthy because natural gas, which is used to produce nearly a third of the country's power, is 43 percent cheaper than a year ago. A *long-term downward trend* in power prices could be starting to reverse… Adjusted for inflation, the average retail electricity price has been drifting mostly lower since 1984." The article did note: "In many states, retail electricity rates are set by regulators every few years. As a result, lower power costs haven't *yet* made their way to customers." *Emphasis added*

The public's perception is to be expected. Headlines about terrifying rate increases are ubiquitous. Electric rates and bills must be high. They must be heading even higher all the time. If someone argued otherwise, they would be met with a blank stare.

Must be true

Everyone says it. Everyone believes it. It must be true.

Utility rates and bills are high. They're very high. And they keep increasing. And they're going to keep increasing a lot.[11]

On April Fool's Day 2011, I read a news headline saying these crazy things in large bold type. It was neither a joke nor unusual. You see this kind of nonsense all the time. In print, it's a great story, guaranteed to sell papers. In electronic media, it's a great story, too, guaranteed to get clicks.

Friends of mine who work for utilities know that these sensationalist claims are untrue, just like I do.[12] Why doesn't anyone ever fight back? Why don't utilities rebut this stuff with the facts?

It must make the public wonder. Why won't utilities dispute this storyline? The only reasonable explanation is that they themselves know it to be true.

Indeed, utilities often unwittingly reinforce the common belief that utility rates and bills are high and constantly heading higher. Just look at what happens when utilities propose rate increases. They seem genuinely apologetic about the added burden on their customers. This must mean that utility rates and bills are high.

---

11 Dennis Cauchon, "Electric Bills Soar $300 in 5 Years," *USA Today*, Dec. 13, 2011. "Electric bills have skyrocketed in the past five years, a sharp reversal from *a quarter-century when Americans enjoyed stable power bills even as they used more electricity*." Despite this opening line of the article and its headline, the article admits: "A potential bright spot: Electric bills appear roughly the same so far this year as last when adjusted for inflation, based on preliminary reports." The article also admits that roughly half the increase in the prior five years was due to greater electricity use by households. Cited is an analysis of Bureau of Economic Analysis data by the newspaper showing that electricity costs consume $1.50 of every $100 of Americans' after-tax income. Some would say that electricity costing just 1.5 percent of after-tax income is an extraordinarily positive achievement. *Emphasis added*

12 American households' average electricity price per kilowatt-hour increased just 1.4 percent from 2011 to 2012, from 11.75 to 11.91 cents, both November year-to-date. For American businesses, prices actually decreased, from 10.28 to 10.14 cents for commercial customers of utilities, a 1.4-percent decrease, and from 6.85 to 6.71 cents for industrial customers, a 2-percent decrease. Though there was a small price increase for households before appropriately adjusting for inflation, real prices for them also fell since the Consumer Price Index went up 2.1 percent from 2011 to 2012. Utility rates are certainly not increasing generally. Source: Energy Information Administration, Bureau of Labor Statistics.

And look at what happens when utilities push energy efficiency. They give the impression that their customers can save serious money by cutting back electricity use. This must also mean that utility rates and bills are high.

It's not only the utilities. From every direction the public hears that utility rates and bills are high and constantly heading higher. Those hawking rooftop solar and home generators say it. Those hawking their environmental positions on both sides of the issues say it. Those hawking their opposition to proposed rate increases say it too. So it must be true. Isn't it?

Down this road

On that early spring day in 2011, I sat back and asked myself, *If I'm so sure this storyline about utility rates and bills is hogwash, why does everyone repeat it? And if I'm so sure the common wisdom is wrong, could I show that definitively?*

That's how I got started. That's when I took the first step down this road to explore electricity's value.

I figured that the storyline about utility rates and bills was so widely accepted that only an extraordinarily compelling series of rebuttals could stand a chance to turn the tide of public opinion. Each rebuttal would need to be rock solid and easily understood. There would have to be several analyses, all pointing in the same direction, that utility rates and bills are actually moderate. There couldn't be one kind of analysis left out, lest doubters and cynics could cry, "See, you didn't look at that."

So I wondered, could I pull this off? Well, what the heck? What's the worst thing that can happen? The public's view of utility rates and bills couldn't get any worse.

Now for Part I

In Chapter 1, "Is It a Bad or Good Deal?" you'll read about how it all started. What drove me to explore the unmapped world of electricitynomics in the first place?

In Chapter 2, "The Value-for-Money Question," you'll see how this research came together. You'll see how we formulated a big question to challenge the team and drive all of our analysis.

The journey began innocently enough, with an itching curiosity about electric rates and bills. Part I will report the story of these first tentative steps.

% Household Expenditures on Electric Bills Unchanged

1984	2011
2.86%	2.86%

Much has changed since 1984. Yet, in that year, when the first "Consumer Expenditure Survey" was conducted, 2.86 percent of the average American household's expenditures were for electric bills. In 2011, the very same 2.86 percent were for electric bills, despite *much* greater use and value of electricity.

This percentage has been in a narrow range of 2.75 percent (plus or minus 0.14 percent) in every year of the twenty-eight–year period, except during 1998–2005, an eight-year sub-period when it was in a narrow range of 2.50 percent (plus or minus 0.11 percent).

Chapter 1: Is It a Bad or Good Deal?

Claims, perceptions about electricity prices; when electricity is free; regulation's extraordinary track record

Legend

Electric rates and bills are high. And they're endlessly increasing.

Come on now. This is an urban legend. Though most people buy into it.

This line, that electricity constantly gets more expensive, is constantly repeated. Say something enough, and people will eventually believe it.

It's popular too. Pronounce electricity as too costly. The audience replies, "Amen." Mitt Romney, the Republican candidate for president in 2012, fell back to this line time and time again.[13]

Pushing your organization's agenda against grid electricity? Throw in that electricity is too costly. It's a proven strategy.

The legend about grid electricity's cost has been popular and effective. But it's actually fairly easy to expose the sham. The chinks in the armor are gaping.

---

13 Calvin Woodward, "Debate Fact Check: How Obama, Romney Statements Hold Up," Oct. 3, 2012, *Huffington Post*. The AP fact-checked Republican candidate Mitt Romney's statement in a presidential debate: "At the same time, gasoline prices have doubled under the president. Electric rates are up." The AP concluded: "Retail electricity prices have risen since Obama took office—*barely*. They've grown by an average of less than 1 percent per year, *less* than the rate of inflation and slower than the historical growth in electricity prices." *Emphasis added*

Here's a preview. Look at how common it is that electricity is given away for free. That's right. Grid electricity is often given away for free.

How odd. Costly things are generally not given away for free. But electricity is.

Does this tip you off? How's that legend about electricity looking now?

No charge

Stop in at a Starbucks or other café. And plug in. Whether it's your laptop, your smartphone, your e-reader, your music player, whatever, go ahead and plug in. The electricity is free.

Suppose you're on the road. Get to your hotel room. And plug in. Use as much electricity as you'd like. Doesn't matter what the hotel pays its utility, at least to you. Electricity's cost never, ever shows up on the tab.

And what about the spots to power up in airports and train stations? On the road, places to plug in are ubiquitous. Again, there's no cost.

It can be true even when you're on the move. Some planes and trains let you power up, gratis of course.

Many businesses like bookstores and health clubs do the same thing. Public buildings like libraries, courthouses, and hospitals offer free electricity too. No questions asked. The times may be tight, but you wouldn't know it from the presence of free electricity all around us.

It's the same story in a college dorm or practically anywhere else on campus. And it's the same thing in workplaces. Plug in a few lamps, a fan, a computer, a printer, and a clock radio. There's never any objection to that. Everybody does it. You can even plug in a TV, a video recorder, a refrigerator, and a microwave oven. There's effectively no limit.

Okay, you're limited by your imagination and by what Best Buy sells. But, really, electricity is a virtually unlimited, free good. Once you step out your door (for college students, once you step out your parents' door), you step into a world of free power. Who among us has ever paid for electricity outside their home?

It's not uncommon for electricity to be included in the rent for an apartment, a condo, or a vacation rental property. Use three hundred kilowatt-hours in a month. Or use three thousand kilowatt-hours in a month, ten times as much, running your electrical equipment hard. It doesn't matter to anyone. It doesn't cost you a penny more to use ten times more power. To economists, that means electricity's *marginal costs* are zero. In other words, the stuff is free.

We can go on and on. You get the picture. Free electricity is the norm outside the home. It's easily available and plentiful.

What does this tell us? What does this say about electricity's costs when somebody actually pays for it?

Born free

Why do they do it? Why do businesses and institutions so freely give away grid electricity?

A watt-hour costs around a hundredth of a cent to whoever is footing the bill. Total up all the freebie energy passed out by a corner Starbucks or even by a major metropolitan airport. It is small potatoes in the scheme of things.

After all, travelers don't tend to lug along with them to the airport their fifty-inch plasma TVs. They don't plug in, sit down, and watch their favorite shows for hours. College freshmen don't tend to install portable air conditioners with dehumidifiers in their college dorm rooms.

Although, that fifty-inch plasma TV uses only so much power. If the TV is on for an hour, utilities charge around three cents for the three hundred watt-hours or so delivered. That's not enough to break the bank for anyone.

The portable air conditioner is an energy hog, in comparison. If the AC is on for an hour, utilities charge around ten cents for the one thousand watt-hours or so delivered. Even this amount isn't all that much. But expect a few raised eyebrows when wheeling an AC unit into a bookstore or library, plugging in for an extra cool breeze.

Blowin' in the wind

Electricity is not only often given away for free. It is often used by businesses for marginal benefit. Lights are left glowing in retail stores at night as cheap advertising. Front entrances are left open, the cold air from air conditioning luring passing shoppers. Florida hotels are frigid from air conditioning gone wild. If grid electricity were expensive, do you think that businesses would do these things? Do you think they would waste so much money for such a marginal benefit?

No way. The benefits are indeed marginal. But businesses know full well that the cost of electricity for these tricks of the trade is small change. It's well worth it.

Free utilities

Utilities would be tempted to do the same, if utility regulators would ever allow them to. It would be tempting to skip metering, billing, and collecting for the drips of electricity taken by so many households, just as many businesses and institutions skip all this for the drops taken by their customers.

Millions of American homes take less than twelve kilowatt-hours daily. Keep in mind this is merely an average. On some days, these households take more; on some, they take less.

Less than twelve kilowatt-hours daily is the same as less than five hundred watts per hour. Again, this, too, is merely an average. In some hours, these households will take more, like in the early evening after work. In some hours, they will take less, like in the middle of the night.

But any way you cut it, less than twelve kilowatt-hours daily, or less than five hundred watts hourly, is a drip of electricity. Just think. If a household takes less than five hundred watts hourly, that's like using less grid electricity than it takes to light five one-hundred-watt incandescent bulbs. No air conditioner. No TV. No washer/dryer or refrigerator or phone charger or...you get the point. Less than five hundred watts hourly is pretty drippy.

The monthly grid take would be no more than 360 kilowatt-hours (thirty days times twelve kilowatt-hours daily). At the rate of a dime per kilowatt-hour, a typical rate set by utility regulators, a monthly bill totals thirty-six dollars. Per day, that comes to $1.20.

During the winter, spring, and fall, when air conditioners may be idle, light-usage households average even less. Their monthly bill may dip to $22.50. Per day, that comes to seventy-five cents.

These customers that take drips rather than profuse flows of grid electricity are typically around one-quarter of a utility's residential customers. That's a real big segment. If a utility serves 2 million households, around 500,000 of them take only drips.

These 500,000 customers may cover just one-twentieth of what the utility collects to operate and maintain its end of the grid. One-twentieth, that's no more than a sliver. Their contribution to what the utility needs to keep up the grid is no more than symbolically important.

Utility regulators could make electricity free to the drippers. There's no way regulators would do this, of course. There's no way they'd allow this kind of "discrimination."

There is an almost sacred doctrine abided by utility regulators. Utility rates must be non-discriminatory rates. Utilities must be made to meter, bill, and collect for the electricity taken by every single customer. Every customer must pay something regardless of how tiny the customer's take.

Nickel daily

A utility proposes that utility regulators approve a hike in electric rates. The announcement drives the media and "consumer advocates" wild with indignation. It's all very dramatic.[14]

Yet what would happen if the utility's proposal were approved? Most residential customers would have to pay about a nickel more daily for their family's electricity. This small amount would devastate few family budgets. But you wouldn't know that from the headlines.

Not for a household that is a heavy user of grid electricity. They would naturally have to pay more. Their impact may be ten to fifteen cents daily. Heavy users typically live in large single-family detached houses and have high incomes.

14 AARP New York, "AARP Stops Electric Price Increase," press release, Aug. 30, 2011. "With the third highest utility prices in the nation, it's no surprise that many New Yorkers are struggling to pay their very high utility bills. In fact, an AARP survey, conducted earlier this year, revealed that 40% of New Yorkers age 50+ have difficulty paying their monthly electric bills." Yet, according to the Energy Information Administration, New York households pay around three dollars daily for electricity on average, those that are low-income pay around $2.50 on average, and those that are low-income and elderly pay around two dollars on average. At these levels, any electric price increase would cost vulnerable New Yorkers a few cents daily.

If a household is a light or medium user, its impact will be in the range of five cents daily. Light and medium users typically live in modest single-family detached houses, single-family attached houses, or multi-family apartment buildings and have low or medium incomes.

The impact on the vulnerable low-income segment of the population would be a nickel daily? That probably isn't dramatic enough to make the headlines, nor to marshal public anger. Instead, the public hears of enormous backbreaking rate increases, forcing families and elders to choose between electricity on the one hand and necessities like food and medicine on the other.

Here's a little secret about these rate increase proposals. When a utility proposes a hike in rates, it knows very well there is virtually no chance regulators will approve it in full. Regulators almost never approve a utility's initial proposal, not even close.

Here's how it works. A utility proposes a rate hike. Regulators could theoretically approve it in full. The utility would then net x dollars more, annually, to operate and maintain its corner of the grid.

But after a year or so of exhausting deliberations, regulators may decide to allow an increase of half of x or a third of x. This is common regulatory practice. So the ultimate impact on utility customers is always overstated in the headlines covering a utility's initial rate hike proposal.

You ask why no one mentions this little secret in the press. Utilities don't want to admit that rate hikes are highly unlikely to be approved in full. Opponents to rate hikes don't want to either.

Consumer advocates would rather use the overstated rate impact numbers. This stirs up public anger. After regulators chop down to size a utility's initial proposal, inevitably, the advocates for minimizing rates can claim credit. Funded by fees tacked on customers' utility bills, the advocates need to justify what they do.

Clearly the media has no interest in revealing the secret. Why mention that, while the utility's initial proposal is x dollars or x percent, regulators will undoubtedly end up at half of x or a third of x? The proposed hike wouldn't seem that newsworthy.

Which is it?

Which is it? Is electricity so affordable for businesses and institutions that it can be given away to customers like paper napkins? Or is electricity so unaffordable that it justifies a sophisticated campaign of opposition every time utilities make proposals to utility regulators?[15]

We get it that these sophisticated campaigns against utility proposals are logical for the bottom line of the businesses and the institutions that sponsor them. We get it that these campaigns are profitable for the interest groups of the businesses and institutions.

Spend a few dollars to poison the well for the utility when the utility goes before its regulator with a proposal. Rally the media and general public, too, against the proposal. Reap the reward. Businesses and institutions will get lower utility bills when utility regulators respond, cutting the utility's proposal accordingly.

The return on investment, for these "interveners" in utility rate proceedings, can be real good. Put some dollars into lawyers, testifying witnesses, and advertisements. Get many more dollars back by forcing down utility bills through regulatory activism.

Those who care that the grid is sufficiently funded to deliver needed power around the clock may have a different perspective. The regulatory activism by businesses and institutions has three gigantic downsides for the country.

First, it directly suppresses funding for the grid likely below the levels that make sense. Second, it indirectly suppresses funding by souring the media and general public about every utility proposal, no matter its merit. Third, and perhaps most importantly long-term, it lends credibility to this line that electricity is unaffordable. This effect also indirectly suppresses funding by souring the media and general public.

There was a time, decades ago, when various groups alleged that utilities had too much funding, enough to "gold-plate" utility facilities. Does anybody seriously believe that utilities have too much funding today?

---

15 Peter Fox-Penner, *Smart Power*, Island Press, 2010. "Most electric regulatory policies have the effect of reducing electricity prices below 'free market' levels. Since underpricing electricity causes people to use more of it, a combination of underpriced power plus publicly funded [energy efficiency] programs is a wasteful contradiction. It is like the government giving out free candy and then running programs for weight loss at the same time. It is indeed regrettable that many utility regulations and pricing policies tend to suppress prices when they should be higher."

Bad joke

The "too cheap to meter" joke is told at utilities' expense. It eviscerates them.

The bad joke goes like this. Utilities promised that their electricity, because of nuclear power, etc., would become too cheap to meter. Isn't that a hoot? Look at how expensive their electricity is. Can we believe anything utilities say?[16]

The joke cherry-picks a speech given some sixty years ago by the chairman of the US Atomic Energy Commission, Lewis Strauss. Strauss, by the way, was a regulator, not a utility representative. Speaking to the National Association of Science Writers, he said at one point, "Our children will enjoy in their homes electrical energy too cheap to meter."

Why did Strauss say this? Remember, Strauss was a regulator. The guy wasn't a utility representative. This wasn't a silly prognostication of utilities. It was an over-the-top statement by a government bureaucrat in the President Eisenhower Administration. The man was pushing hard for more research into fusion nuclear power. The line didn't have anything to do with the prospects of fission nuclear power, which is what utilities eventually went with.

Utility representatives never promised that the electricity they deliver to Americans would become too cheap to meter. Why would they? Utilities are businesses that charge customers based on what the meter says. So the joke isn't just bad. It's dumb.

While we don't enjoy in our homes electrical energy too cheap to meter, we certainly do enjoy electricity that is almost that economical.[17, 18] Even Jules Verne couldn't have conjured up a better prediction.

---

16 Joseph P. Tomain, "'Steel in the Ground': Greening the Grid with the iUtility," *Environmental Law*, 2009. This is just one of numerous references to the misleading "too cheap to meter" quote.

17 American households' electricity costs as a percentage of their total expenditures were identical in 2011 to what they were in 1984, twenty-seven years ago. Grid electricity took 2.86 percent of households' expenditures in both 2011 and 1984. This is a remarkable accomplishment on the part of utilities and their regulators, especially because electricity usage per household has grown considerably during the last three decades, as has utilities' costs for environmental quality. Source: Bureau of Labor Statistics, Consumer Expenditure Survey.

18 Outside of the South, American households' electricity costs, as a percentage of their total expenditures, are particularly low. For households in the West, the average is just 2.1 percent. That's about one-fiftieth of their expenditures. These households actually spend more on telephone services than they do on grid electricity. For households in the Northeast and Midwest, the averages are slightly higher, at 2.5 percent and 2.6 percent, respectively. Source: Bureau of Labor Statistics, Consumer Expenditure Survey.

Bill increases in households' budgets

Consider a low-income household. It has an electric bill of $900 annually. This equals $75 monthly, $2.50 daily.

Its electricity cost is higher than $75 in summer months due to greater use of electricity for air conditioning. Its cost is lower during the other seasons.

Suppose the utility proposes a rate increase to utility regulators. Headlines scream out the terrifying news. The proposed increase is in the hundreds of millions of dollars. The typical customer would bear 10 percent higher bills.

The media might neglect to mention that the 10-percent hike would affect the transmission and distribution portion of electric bills only. The sizable power plant portion would not be affected (nor portions for taxes, nor for state-mandated fees).

The transmission and distribution portion is generally around half of electric bills. Then the proposed 10-percent hike would impact households' bills by half of 10 percent, or 5 percent.

Suppose the utility won't propose another rate increase for three years. Over this three-year period, a 5-percent increase is equivalent to an increase *per annum* of 1.7 percent (without compounding).

Suppose inflation averages 2 percent per annum for each of the three years, totaling 6 percent over those three years (without compounding). This inflation transforms a 1.7-percent per annum increase into a 0.3-percent per annum decrease.

Utility regulators rarely allow all that utilities propose. Suppose that half of the utility's proposal is allowed by regulators.

The increase would actually be 2.5 percent over the three years, unadjusted, and therefore 0.8 percent *per annum*, also unadjusted. Inflation transforms a 0.8 per annum increase into a 1.2 percent per annum decrease. The increase would not be an increase at all.

The per annum impact on the low-income household, unadjusted, would be around sixty cents monthly. That's two cents daily. Because utility regulators increase rates every few years and keep them constant in the interim, you have to look at all of the years, not just those in which rates increase.

By the end of the three-year period, its average bill would rise to $76.80 from the current level, $75, an increase of $1.80 (unadjusted for inflation and rounded). But this $76.80 would feel like $72.30 to the household in inflation-adjusted "constant dollars" (rounded and without compounding). The decrease was $2.70 (adjusted for inflation). Electricity would have become cheaper than it was.

A utility's proposal to increase rates by 10 percent, as reported by the media, turns out to be a 1.2 percent *per annum* rate decrease. What a difference. When we take into account the portions of electric bills, the multiple years between rate increases, the tendency for regulators to allow only a percentage of utility proposals, and inflation, it's a decrease, not an increase. Unsurprisingly, the headline did not scream out that the proposed rate increase is really a decrease.

Incredible

It's an understatement. Most Americans would be amazed to hear electric rates and bills have been moderate. Yet this is exactly what my research team has proven.[19]

Our seemingly incredible conclusion contradicts the drumbeat of media headlines. It conflicts with what Americans can see for themselves in their own monthly bills. How is it possible that the public perception can be so different from the facts?

Much of what the public hears about electric rates are sound bites. Utility opponents naturally want to pummel a rate increase proposal. It's an effective strategy to paint a vivid, alarming picture of a rate increase proposal. And it's the way to attract media attention and ire.

The media effortlessly churns out tear-jerking, heart-wrenching stories. Through-the-roof rate increases break the backs of already economically strained low- and middle-income families. How many in the public see through the fog of opponent tactics and media dramatizing?

Seeing their own monthly bills, most Americans have no doubts that electricity is expensive and becoming more so all the time. After all, their monthly bill varies wildly from month to month. Bills occasionally jump way up suddenly, for no apparent reason.

19 American households' electricity costs as a percentage of their total expenditures are particularly low for the 20 percent of households with the highest income. Just 2 percent of their expenditures are spent on utility bills. For the 20 percent of households with the lowest income, 4.5 percent of their expenditures are spent on utility bills. That's significantly greater but still moderate, about one-twenty-second of their expenditures. Source: Bureau of Labor Statistics, Consumer Expenditure Survey.

There is constant media chatter reinforcing the common perception. Politicians, pundits, and promoters of competing services put out that electric rates and bills are high and heading higher. Overzealous solar-panel advocates and home-improvement contractors can make intemperate, unsupported claims about electric rates and bills all they want.

Media disconnect

In May 2011, *Yahoo! News* posted a provocative AARP Illinois press release: "AARP, Groups, Urge Lawmakers to Stop Utility Rate Hike Express." The piece included a reckless claim that "automatic rate increases will add an unfair and unwarranted burden on the wallets of all consumers, but particularly on those of limited or fixed income among the 50 plus population, who pay an average of 14 percent of their household income on utilities."[20]

Can it be true? The AARP claims that Illinois households comprised of older individuals spend 14 percent of their income on utilities? Inconsistent with national data, it seems incredible. Electric bills are generally around 2 to 4 percent of household budgets.[21]

It's crazy that the average for these Illinois households could be 14 percent. A large number of them must pay more than 14 percent, if the AARP claim is true, and some much more than 14 percent. A few households would be up at 20 percent. This can't be.

Low-income households in the Land of Lincoln, regardless of age, pay only around nine hundred dollars per year on electric bills. Illinois has one of the best track records among the states, keeping electric bills moderate over the last half-century.

Did AARP get to the 14 percent by including a wide range of utilities, not just electric but also gas, water, and communications? Did AARP get to the 14 percent by neglecting the fact that many households with older individuals have far higher household expenditures than incomes because they receive substantial transfer payments such as Social Security?

When the costs of necessities increase, as they constantly do, there is an impact. But we shouldn't sanction the overstating of the impact when the costs of electricity increase.

---

20 AARP Illinois, "AARP, Groups, Urge Lawmakers to Stop Utility Rate Hike Express," press release, May 4, 2011.

21 During the twelve-month period ending July 18, 2012, my parents took 2,774 kilowatt-hours of electricity from the grid for their condo apartment in New Jersey, an average of 7.6 kilowatt-hours daily, at a cost to them of around a dollar daily. This low electricity usage and cost is fairly common among American households.

Consider the case in which electric rates are raised by as much as 5 percent (excluding power plant output costs beyond a utility's control). Suppose a low-income household's electricity costs are $75 monthly, $2.50 daily. The 5-percent increase would come to $1.88 monthly, six cents daily.

Any additional cost on any low-income household is undesirable, whether it has older individuals or not. But six cents daily doesn't justify AARP's inflammatory rhetoric.

Dumb luck

There is a forbidden subject in the electricity profession. The topic is vitally important. But energy experts never broach it.

Utility rates and bills have remained moderate for a half-century. Okay, but how could it have happened? Was it dumb luck? If not that, whom can we thank?

Most Americans would find these questions peculiar. Utility rates and bills have been moderate, that can't be right. Moderate for a half-century—that can't be right either.

You know the common view. Utilities insatiably demand massive rate increases. Their greed is sated with little trouble. Nobody can understand why utility regulators allow utilities such free rein. A plausible explanation is that regulators are in the pocket of the utilities.

This book lays bare these myths. But that's not enough. Abandoning myth for truth naturally brings up new questions. The half-century track record of Americans' electricity costs is a fact and a feat. How were utility rates and bills held down over the years? It was a feat almost impossible to pull off.

OPEC price spikes in the 1970s struck our fleet of oil-burning power plants. President Carter's Administration then lashed out at natural gas-burning power plants, making gas-burning extremely difficult. Massive cost overruns in the 1980s shocked our construction of nuclear power plants. Clean Air Act amendments in the 1970s and 1990s forced expensive retrofits to our coal-burning power plants to cleanse the air. Each wave meant new grid costs. Utilities and their regulators struggled to squeeze more and more into rates and bills.

We've had escalating commodity and labor costs. Taxes have been tacked on. Fees have been tacked on to fund the favorite initiatives of political leaders. Each escalation, tax, and fee meant new grid costs. Regulators and utilities struggled further to squeeze everything into rates and bills.

Americans take more electricity from the grid than ever. That's right. We take more from the grid than even a decade ago. What about those feel-good stories in the media about households and businesses "going green"? Government data show that they are a distinct minority.

If per-kilowatt-hour rates are rising, and what we take from the grid is rising too, then utility bills should be soaring. But they're not. Dare we ask what we're doing right here?

There's a simple answer, one that is guaranteed to madden energy experts. They're too close to the subject to have perspective. Sorry, experts. Depending upon your agendas, you'd like me to say that the answer is environmental policies or that the answer is restraining environmental policies. Depending upon your agenda, you'd like me to say that the answer is overbearing utility regulators or that the answer is regulators who let utilities do their job. I just can't.

The answer, instead, is a seldom-lauded invention of early twentieth-century America. Utility regulation, the shotgun marriage of utility regulators and utilities, is what has kept utility rates and bills down through all of these trials.

Utility regulation faced down big challenges. It dispatched waves of new grid costs all while Americans took more and more from the grid. Through all of this, utility regulation kept electricity to less than 3 percent of average household expenditures. It was a remarkable achievement, not dumb luck.

Reddy's Recharges

1. Grid electricity is often given away for free and used as if it were free, indicating that electricity costs are moderate.

2. The hype surrounding utility rate increases is one reason the public believes that utility rates and bills are high and increasing.

3. Electricity rates and bills have instead remained moderate for half a century, amid waves of cost pressures, under utility regulation.

"I've been selected to pay another electricity bill."

Chapter 2: The Value-for-Money Question

Grid electricity too expensive or too cheap? Defining its value; loss in value from outages

What kind of question is that?

Is it worth it? Is it worth what we pay for it? Are we getting a fair deal? Or are we getting ripped off?

We ask ourselves these questions all the time. Consciously or subconsciously, we ask these questions about every purchase we make. When we buy any good or service, we wonder about whether the seller is taking advantage of us. We wonder if the seller, or a competing seller, can offer a lower price while not compromising quality.

We should be asking these very same questions about our purchases of grid electricity. But we don't. You might ask, why not?

Electricity is too politicized. That's why not. Electricity is rarely seen as a good or service. It's seen instead as a public policy matter.

Everybody from the president to the postman frets about the nation's "energy problems." And there are energy problems. Some naturally command our attention. Take the excessive importation of foreign oil. That's definitely a problem.

Yet real concerns about foreign oil somehow morph into concerns about domestic grid electricity. Both oil and electricity are tossed into the same basket as one and the same.

Electricity is said to have "energy problems" as serious as oil. So we talk about "energy solutions" to fix the problems. We talk about the development of the exciting new energy economy that's coming. It's exciting. It's new.

Few are inclined to talk about the upkeep of the boring old grid that we depend on today. It's boring. It's old.

Fewer still want to talk about whether Americans get value for money. Or could Americans get more value from that boring old grid?

It's weird. A value question seems strangely out of place. One even feels awkward bringing it up.

If you asked about electricity's value to consumers, many energy experts would be, well, flabbergasted. They might frown and shake their heads. How can you ask such a trite question? What audacity! You act as if electricity is like a consumer good or service!

Anybody knows electricity is different. Everybody knows the planet is at stake.

That may be their reaction. It's considered almost immoral in proper company to evaluate electricity as if it was something ordinary like milk or bread. But suppose we did. Just suppose.

What if the grid's electricity were free?

Here's a thought experiment for you. What if grid electricity were free? Not just in your hotel room, not just at Starbucks, not just in the library, but what if electricity were free all the time, everywhere?

How would the world be different? Would we light more and more often? Perhaps every mile of every road would be illuminated like daytime, not only the most-used roads, not just dimly.

Would we air condition more and more often? Perhaps every nook and cranny of every home and business would be cooled like refrigerators, not only the most-used interior spaces, not just a little.

You get the picture. The world in which electricity is free is not that different from our world. It's not some science fiction story. In our world, electricity is almost free even when it's not completely free.

You can tell because we act as if it is free. We use electricity freely. We're quick to invent and adopt new ways to use power, mindless of electricity's cost.

The evidence is overwhelming. Electricity costs little. Think about it this way. Compare electricity's cost to its value in partnership with a machine, appliance, or device you buy. You certainly give thought to the price of the machine, appliance, or device. You hardly give a thought to the cost of the electricity that it will use, despite the energy efficiency label.

Why? The reason actually makes a lot of sense. The cost of electricity that the machine, appliance, or device will use will be small and spread thinly over years. So we give it hardly a thought.

Here's another way to see that electricity is almost free. Listen to those who dearly want Americans to use less electricity. This crowd is not coy. They feel electricity's cost is too close to free. They decry this as a societal sin. They feel electricity's cost should be way higher to reflect its environmental impact. They want government to make the cost way higher.

Dollar a day

Most low-income households, except in the South, pay around two dollars for their electricity per day on average. Leaving aside the steamy summer months when air conditioners run hard, these households pay around a dollar a day on average. That's not free electricity, but it sure is close to free!

You have to laugh when you hear critics say that electricity is too expensive. Do they really believe that a dollar a day is too expensive for these households? Do they think utilities and their regulators should force down further what these households pay? How about seventy-five cents a day? Would that be right? How about fifty cents a day? Would that be right?

At some point doesn't it become ridiculous? Why charge these households anything for their electricity? Let's not mess around. Let's just go ahead and make electricity free for these households for the winter, spring, and fall, and perhaps the summer too. That would be silly, of course. But it shows how little electricity's cost is.

Imagine if electricity were expensive

The studies have been definitive. Utility rates would have to go way up. Utility regulators would have to jack rates up as they have never done before.

Otherwise, most Americans won't budge. They won't make wholesale cutbacks in electricity usage. They certainly won't cut usage as dramatically as is hoped by energy efficiency boosters. It just won't happen given how low utility rates are.

That derisive joke about grid electricity cites an optimistic quote in 1953, out of context, that electricity will become too cheap to meter. Electricity didn't become that cheap. Utilities do meter what your grid take is.

Electricity isn't too cheap to meter but too cheap to matter. You probably don't spend much time looking to cut a watt here and a watt there. Few Americans do. It only makes sense. Why spend a lot of your time sweating ways to save a few pennies? Most people have better things to do.

If regulators doubled or tripled or quadrupled utility rates, you'd care more about saving watts. You'd spend some time considering your options. You'd be more inclined to make major changes to your lifestyle if there were real monetary savings to be had.

But, given how cheap electricity is, changes in your lifestyle to cut watts is one of the least-rewarding ways to save a buck or make a buck. You can do better selling your old lava lamp on eBay.

I'm leaving aside those really enthused about reducing their carbon footprint. They're ready to remake their lifestyle to save some watts. But those Americans are a minority.

For most Americans, how much would regulators need to jack up utility rates so that electricity's cost begins to matter? If utility rates doubled or tripled or quadrupled, would that significantly change consumer behavior? That's hard to say. It's never happened.

Utility rates have generally decreased or remained relatively stable in real terms throughout history. This has been true since Thomas Edison's day.

Utility regulators have never doubled utility rates, let alone tripled or quadrupled them. Doing so would be unthinkable, even for a cause seen as noble and popular as spurring utility customers to cut back electricity usage.

However, we can find isolated instances in which utility rates jumped temporarily by a high percentage. In these rare cases, utility regulators were unwilling bystanders to forces outside their control.

Set aside for now the circumstances behind these cases. These unique situations enable us to ask a gigantic question. How did utility customers react? What did consumers do when electricity's costs went from not mattering to heights impossible to ignore?

Too expensive and too cheap

A range of special-interest organizations work quite hard to shake up and reshape the way electricity is delivered and used. These groups cajole, coerce, and compel regulators to impose "energy solutions" to solve perceived "energy problems."

For them, electricity's value question is irrelevant. Its answer is obvious. Grid electricity is a bad value for us all, every single household and business. Grid electricity, as they see it, is indelibly linked to the threat of environmental degradation.[22] It's impossible for them to consider grid electricity as a highly valued good or service with ordinary economic characteristics.

For them, grid electricity is both too expensive and too cheap. They seem convinced that a new breed of twenty-first-century green-energy entrepreneurs will bestow on consumers less costly electricity. In the very same breath, they seem convinced that future consumers will use electricity less. Which is it, guys? You can't make something cheaper *and* used less. Well, you can't without elaborate government intervention into supplier and consumer behavior.

22 Seth Borenstein and Jonathan Fahey, "Officials: CO2 Emissions in US Drop Because of Plentiful Natural Gas," *Scranton Times Tribune*, Aug. 16, 2012. "The International Energy Agency said the U.S. has *cut* carbon dioxide emissions *more* than any other country over the last six years. Total U.S. carbon emissions from energy consumption peaked at 6 billion metric tons in 2007. Projections for this year are around 5.2 billion, and the 1990 figure was about 5 billion." *Emphasis added*

Critic confusion

Critics slam grid electricity from opposite directions. Sometimes it's that electricity is too costly; utility rates need to go lower to protect people. Sometimes it's that electricity isn't costly enough; utility rates need to go higher to protect the environment. The critic confusion is schizophrenic. Not surprisingly, the public is totally confused about electricity's cost and value. That is, those in the general public not completely turned off by all the back and forth.

The one conclusion that must be clear to the public is that there are real problems in electricity. The public may not have the patience to sort out the too-costly slam from the not-costly-enough slam. But all of the negativism makes people convinced something is wrong with what utilities and utility regulators have been doing.

If your special-interest group is pushing for less-costly electricity, that there's free and nearly free electricity throughout our society is embarrassing. If your special-interest group is pushing for more-costly electricity, that there's free and nearly free electricity is disgusting. Neither kind of special-interest group wants to acknowledge electricity's low cost. It's a no-no.

Difficult question

This question has stumped energy experts for decades. How do utility customers value grid electricity?

The question doesn't seem difficult, at first. The question also doesn't seem practical. What does it matter? We all know that electricity is valuable. We know it's valuable when we think about it, which is rarely. So what?

Electricity's value question is actually uniquely difficult. The contrast with most goods and services is stark. For them, the value question isn't that hard.

Freely set prices are the key. The prices of most goods and services approximate the value to consumers.

Movie ticket prices rise. Many consumers stop going to theaters as often. Many stop going completely.

Look what happens when movie ticket prices rise from ten to fifteen dollars. Those consumers who cut back or cut out theatergoing reveal by their reaction their valuation. For them, a ticket has a value somewhere in that very same range of ten to fifteen dollars.

But grid electricity doesn't have freely set prices. Utility regulators set electricity prices. And they set prices without any regard to electricity's value to consumers.

Instead, regulators rely on arcane complex formulas. It is all hashed out in epic litigious clashes of regulatory contenders. Value is rarely considered or mentioned during these turgid procedures.

How utility customers value grid electricity is a difficult question for other reasons too. Electricity has unusual characteristics among all goods and services. These characteristics make electricity's value elusive.

Electricity's value, for one thing, is mainly apparent when its continuous provision by the sellers (the utilities) is interrupted. How strange. We're aware of our consumption of most goods and services when sellers provide it to us and when we make use of it. This isn't true of electricity.

Ask a friend how he values electricity. He'll probably find your question vexing. "Who knows? Now, can we turn the game back on?"

Wait 'til the power goes out. It was a bad storm. The winds pulled down trees and wood utility poles with them.

Ask your friend again. How do you value electricity now? Guaranteed, your question will seem clearer, more meaningful. Suffocating from the lack of power, he'll come up with a pretty good dollar number.

Weird powers

Why is electricity so weird? There are several factors at work here. Electricity is a necessity. Sure, it's not up there like food and water. But is it that far below them in modern America?

Electricity has no close substitutes. Without bank ATMs, store checkout machines, gas station pumps, smart phones, lights, air conditioners, and on and on, your world is put on hold without any recourse.

Grid electricity, however, has growing competition from pretenders. Purveyors and promoters claim solar roofs, natural gas generators, and micro-grids are adequate, close substitutes for grid electricity. We'll see later in this book how the claims come out.

Electricity is used constantly. There are few times when you're not taking watts from the grid yourself or when someone isn't taking watts on your behalf. Think about it. When's the last time you went without the grid? When walking the dog in the park? Bet your smart phone was in your pocket. Bet the officer on duty you just passed has electric communications on him. Bet the water fountain depends on electric pumps.

Very few goods and services have these same characteristics. Police and fire department services and national defense are necessities just like electricity. They, too, are used constantly and mindlessly. But, unlike electricity, their continuous provision by sellers is never interrupted, thankfully.

Communications services are necessities just like electricity. They, too, are used constantly and mindlessly. But, unlike electricity, their continuous provision by sellers is rarely interrupted by lengthy outages. When their continuous provision is interrupted, "the system" is always down for brief periods of minutes, a couple of hours at the most.

We cannot say that this is the case for the provision by sellers of electricity. Regrettably, its provision is interrupted from time to time. The grid, in what we aptly call a blackout, can be down for days or weeks.

Strict price controls

Answering electricity's value question is essential. If electricity's price (per-kilowatt-hour rates) approximates its value to households then utility regulators should resist raising the price. Doing so would force households to cut back what they take from the grid. This result might please energy efficiency boosters. But it would translate into less comfort, convenience, and entertainment for households and hit businesses hard.

Suppose that, instead, electricity's price is far less than its value to households and businesses. That's not so hard to imagine. After all, utility regulators have been maintaining strict price controls for nearly a century on what utilities sell. If electricity's price was far less than its value then regulators shouldn't resist raising the price for worthy reasons. So it matters a lot whether or not electricity's price is far less than its value.

All of the evidence is that electricity's value to households and businesses greatly exceeds its cost to them. Indeed, the value gap has been rapidly increasing over the last two decades.[23]

Defining value

Suppose that grid electricity had no meters, that utility regulators mandate it in one sweeping order. Take as much as you want. No per-kilowatt-hour charge ever!

Follow this fantasy further. Suppose that grid electricity is always there for you. It's like your best friend. The grid never lets you down. Power you can count on. No blackouts, ever.

You get all the electricity you want. There's never a blackout. What would that be worth to you? For a day, would it be worth ten dollars? That's what you might be spending daily on your commute to work. Would it be worth fifty dollars? That's what you might be spending daily on your mortgage or apartment rent. Would it be worth somewhere in between?

Wait. Don't answer yet. You need to consider the question's converse at the very same time.

You don't get all of the electricity you want. Actually, you don't get any electricity. Nor do your neighbors or community. As in a blackout, without power, it's dark and lifeless.

Put this way, the question makes electricity more like other goods and services. You either buy it or you don't. You either get it or you don't.

No electricity for you, your neighbors, your community. What would that cost you? Would it cost you ten dollars? Would it cost you fifty dollars? Would it cost somewhere in between?

It must be obvious by now. It's not an easy question to answer, whether posed directly or in the converse.

---

23 The average American household spent 2.82 times as much on telephone services in 2011 than in 1984, but only 2.26 times as much on electricity. Inflation was at work—the average household also spent 2.26 times as much on everything totaled. So over the last three decades, the net value of both telephone services and electricity has rapidly increased. Source: Bureau of Labor Statistics, Consumer Expenditure Survey.

Despite its complexity, electricity's value question offers real lessons. If grid electricity were cost- and blackout-free, how much would that be worth to you? "Perfect power," electricity free of any costs or worries, would be really valuable.

Nothing but net

Grid electricity is, however, not perfect. It's not cost-free. As economical as the grid is, utility bills are some number of dollars and cents. It's not blackout-free. As reliable as the grid is, utility blackouts are some number of dollars and cents, too, if you think about it.

Every time your utility has a blackout, you incur a cost. The longer the blackout, the greater your cost. Indeed, you incur a cost merely knowing that utility blackouts will take place from time to time and that they'll occasionally be long.

Utility bills cut into electricity's value. The higher utility rates are, the lesser electricity's value is to you.

Utility blackouts also cut into electricity's value. The more frequent and longer utility blackouts are, the lesser electricity's value is to you.

The value of grid electricity to households and businesses is thus a net quantity. It is net of the costs of your utility bills and your blackout-related costs. Both squeeze your net value.

Go back to the value question. You can still get all of the electricity you want. But now there's a charge, and there's the occasional blackout. What would avoiding blackouts be worth to you? For a day, would it be worth ten dollars? Would it be worth fifty? Would it be worth somewhere in between?

For you, utility bills might be three dollars daily. Avoiding blackouts might be worth fifteen dollars daily. That's five times more than what you pay in utility bills. How can we arrive at a number like fifteen dollars for blackouts?

We include all the costs of blackouts and spread them evenly over all the days of the year. We include the out-of-pocket expenses of blackouts. We include the inconvenience of blackouts. We include the health and safety risks of blackouts. We include the lost work productivity of blackouts. And we include the mental worry of blackouts.

If meter-free, blackout-free, perfect power is worth thirty-five dollars daily to you, the net value of electricity is seventeen dollars daily. We simply subtracted from perfect power's value, thirty-five dollars, both the cost of bills, three dollars, and of blackouts, fifteen dollars.

So suppose that the value of electricity, net, was seventeen dollars daily. That's nearly six times more than your utility bills. Nice!

Regulator magic

As a utility regulator, you have to feel pretty good about the situation. Every single day you're giving a great gift to all your constituents—your state's businesses and households. You're charging households something like three dollars but giving them something worth thirty-five dollars, if it was perfect, and twenty dollars after its imperfection is factored in. That's value for money. You're a rock star. Your constituents should be clamoring for your autograph.

Here's the really exciting part of all this. As good as the present deal is, you can make it even better for your constituents. You can jimmy with your formula a little bit to create and bestow on them even more value.

As a utility regulator, you have it within your vast powers. You may not feel all that powerful, but you actually are. Regulators' modesty is an admirable trait. But give me a regulator who is always looking for ways to maximize how much the public can benefit from grid electricity.

Look at the millions of households served by the utilities you regulate. You can engineer a tradeoff. Trade some bill costs. That's easy. Regulators routinely make households' utility bills higher or lower when they set utilities' per-kilowatt-hour rates. Get in return some blackout costs. That's not that hard either. Regulators routinely order utilities to develop well-thought-out plans to spend money on this or that for the public good.

You can do all that. You can literally trade pennies on the dollar. Just like that.

You simply order your utilities to cut households' blackout costs by a good amount. Your utilities can do that. We know they can, as long as they have the money to do it with. If what it would cost in households' bills is a lot less than what it would save in households' blackouts, you're there. Just wave your regulator's wand. It happens. It's regulator magic.

You can do all that for your state's businesses, as well as its households. While as much as your state's households incur a heavy price for utility blackouts, your state's businesses may incur a heavier price. Can your utilities save businesses' blackout costs, swamping the needed increment in businesses' bills? You bet they can.

Sweet

It was nice being a utility regulator, if only for a little while, wasn't it? Now switch back to your role as a plain old utility customer. Remember what that was like? Your utility bill costs were three dollars daily. Your utility blackout costs were fifteen dollars daily.

Total that up. You lose eighteen dollars of value daily from what bill-free, blackout-free, perfect power would be worth to you.

Can regulators and their utilities put their heads together on this? Can they figure out how to reduce this eighteen-dollar loss in your value?

Suppose regulators and utilities figure out how to cut your blackout costs by, let's say, 10 percent. They need to up your bill costs by 10 percent to pull it off.

At first you might protest. Up my bill costs, what nerve! But hold your horses. Let this all sink in. It would be awesome if they could actually do this for you.

Your blackout costs drop by $1.50 daily. That's a lot. Your bill costs rise by thirty cents daily. Okay, that's not a lot. Overall, you make out big-time. Sweet!

The eighteen-dollar loss each day in value falls to $16.80. Let's see what that does to the value you get from the grid. Recall that perfect power, before your losses in value from utility bills and utility blackouts, is worth thirty-five dollars each day. So what was a daily net value of electricity of seventeen dollars is now $18.20.

Your utility bills *did* go up by 10 percent. But the value you're getting from grid electricity, even after counting the bill increase, is up by over 7 percent. It's as if the manufacturer of your favorite breakfast cereal started making the box much bigger while charging a nominally higher price. You're fine with that. You'd be fine, too, if the state's regulators and your utility did the same. The much bigger cereal box more than makes up for a nominal price increase.

Certain uncertainties

Regulators and utilities could, should, devise the most promising set of strategies to cut the public's costs from utility blackouts. There's a catch, though. They can never be sure how much blackout costs they'll be cutting.

How much of these costs will they be cutting by hardening overhead lines, under-grounding select lines, adding line and equipment redundancy, hiring linemen, engineers, and construction personnel, etc.? It can't be known with any great accuracy. How much of utility bill costs will they be adding in the process? This is known exactly.

Regulators, should you throw up your hands at the challenge? You can't know how much blackout costs you'll be cutting for utility customers. But you can and do know exactly how much bill costs you'll be adding to utility customers. Does this make it impossible for you to act to do the right thing for the public? No way!

In the earlier example, blackout costs fell five times more than how much bill costs rose. So what if there's uncertainty as to how much blackout costs decrease?

If we perfectly forecasted how much blackout costs decrease, then the public's benefit cost ratio will be five to one. Suppose we got it wrong. Suppose blackout costs decrease by less than we forecasted. That just means the public's benefit cost ratio might turn out to be something like four to one or three to one.

Three to one, that's still a really beneficial ratio. The value that households and businesses get from grid electricity still jumped by a whole lot.

Suppose we got it wrong in the opposite direction. Suppose blackout costs decrease by more than we forecasted. That means the public's benefit cost ratio might turn out to be something like six to one or seven to one.

Seven to one, that's a super-beneficial ratio. The value that households and businesses get from grid electricity soared like a slam-dunking NBA star.

Given the enormous potential to trade off the public's utility bill costs for utility blackout costs, it's only a matter of time before people demand that regulators act. Nobody likes leaving money on the table in a deal. In the "deal" between utilities and the public, the public's money in terms of excessive blackout costs is sitting right there.

Reddy's Recharges

1. Grid electricity's value question—do households and businesses get a good deal from the grid?—is rarely addressed.

2. Most low-income households outside of the South pay around a dollar a day for what they take from the grid during most of the year.

3. Grid electricity's value—net of utility bill cost and utility blackout cost—can be increased by reducing power outages.

"It saves energy and makes me feel holier."

How We *Pay* for Grid Electricity

On why grid electricity is not expensive despite the contrary perceptions and how its affordability is best understood and communicated

PART II

Part II. How We *Pay* for Grid Electricity

Chapter 3: Prices and Bills

Chapter 4: Breaking It Down, Daily

Chapter 5: Broad Variation in Bills and Use

Proof

I felt confident that grid electricity's cost has been moderate and stable. But I had never delved into it. Though an energy expert for three decades, I had never seen a comprehensive analysis one way or the other.

It seemed odd. Questions about electricity's cost, and its value, are critical to the electricity industry and the country. We hear all the time the claims that electricity's cost has been anything but moderate and stable but with little hard evidence. Why hadn't I seen a good study from any side of the issue?

So I began a search. Were there any good studies out there that I simply missed?

There are studies that include analyses of electricity's cost. But these works are generally narrowly focused on a single utility or a single state. And they tend to be openly biased. They leave no doubt as to whether a particular regulatory action is being advocated or opposed.

I'd need to do my own study. I'd need to design compelling and easy-to-understand analyses using credible public data.

There was no other way to prove my hypothesis that grid electricity costs have been moderate and stable. I didn't have the option to lean on good studies already out there. The only option was to develop the proof myself. That's what I did.

Bulletproof

My proof needed to be bulletproof. Critics would mercilessly try to tear it apart.

Critics might not find it that difficult. A new study asserts grid electricity costs are moderate and stable? Ha! Many Americans would scratch their heads hearing that. Critics wouldn't have to work all that hard. They could merely cast a few doubts. So my proof must not allow any room for doubt.

My proof should consist of multiple proofs. I should address electricity costs in multiple parts of the country plus the nation overall. I should address multiple time periods. No critic should be able to claim that the study worked around an obvious analysis. That if an obvious analysis had been included in the study, its results would contradict everything else in there.

Always the same answer

So the study examined trends in residential electric rates and bills deploying a wide variety of rigorous quantitative analyses. It looked at many time periods and many geographical and demographical cuts of accepted data sets.

It looked at how stable residential electric rates have been, nationally, during the last half-century, appropriately adjusted for general inflation. The finding is always the same. Over the last half-century, rates have usually decreased. Rates have rarely increased. In the instances where rates increased, the upticks were generally well less than the downticks.

Utility opponent groups endeavor to make rate increase proposals seem much larger than they actually are. You can't fault them. Opponents are only doing their job, to throw a scare about utility proposals.

Opponents love to point at the biggest number, the hundreds of millions of dollars at issue in a rate increase proposal. The number is that big, they fail to note, because it's aggregated over all utility customers. The number is that big because it's aggregated over a whole year. The number is that big because all of the utility's diverse needs are combined into one package.

That's their right. Opponent lawyers and consultants are entitled to make their best case why their arguments should prevail in utility regulatory proceedings. But expressing utility bills and proposed rate increases in such terms, as they do, distorts what households actually pay for grid electricity and what they'll pay going forward for rate increases. It also incites the public's fear and anger at utilities, which is the point, I suppose.

Electric days

As I worked on the study, I felt again and again compelled to express electricity usage and bills in daily terms. Per day seemed to be the best way. How many kilowatt-hours does a household take per day? How much does a household pay per day for what it takes?

The daily perspective struck me as easier for the non-expert to understand. Energy experts like me usually talk about usage and bills per month and per year. The scale of what we're talking about is fuzzy. A month is a long time for a consumer who buys many goods and services daily. A year is a very long time.

How many goods and services do consumers buy once per month? How many goods and services do consumers buy once per year?

If you paid once monthly for an entire month's supply of bread for your family, wouldn't the dollar and cents amount sound high? If you paid once annually for an entire year's supply of bread for your family, wouldn't the dollar and cents amount sound really high?

So that's how it began. From that point on, I started expressing electricity usage and bills in easy-to-grasp daily terms. Increasingly, energy experts have adopted this convention. Per day is becoming the way.

Abnormal

The per day thing was a great discovery that just happened as the report was unfolding. It's proven to be a compelling convention. Then a remarkable thing occurred. I made another discovery as the report was wrapping up. This one would have monumental importance. The electricity business, now that we know that so much about it is abnormal, will never be the same.

You would expect that the statistical distribution of households' electricity usage is normal, graphically like the classic bell-shaped curve. You'd be wrong. You would expect that the statistical distribution of households' utility bills is normal. You'd be wrong again, dead wrong.

So much about the electricity business is actually quite abnormal, statistically. The implications of this discovery turn the business upside down.

Now for Part II

In Chapter 3, "Prices and Bills," you'll read about how we started out real basic. If grid electricity's been moderately priced or expensive, it's been moderately priced or expensive compared to what? We needed benchmarks and found good ones all around. We did the same with grid electricity's price stability over time, or lack thereof.

In Chapter 4, "Breaking It Down, Daily," you'll see how this research led to an incredibly simple but just as incredibly powerful unitizing of electricity usage and costs. The team, by expressing kilowatt-hours and dollars and cents in per-day amounts, could now perform analytic feats heretofore confounding to even seasoned energy experts.

In Chapter 5, "Broad Variation in Bills and Use," you'll learn how we happened upon the research's greatest triumph. When digging into how much grid electricity is taken by utility customers of all types, how much they pay, and how much value they realize, we glimpsed the breakthrough for the first time. We discovered the classic normal statistical distribution didn't fit. Something was wrong. Those averages that utilities and utility regulators so depend on, they don't give the full picture. Was there a statistical model that did?

The journey began innocently enough with an itching curiosity about electric rates and bills. Part I reported the story of these first tentative steps. Soon after, the journey was transformed into a wondrous tour full of analytic surprises. The team had its first inklings about electricity's value. Part II will recite the story of these surer steps.

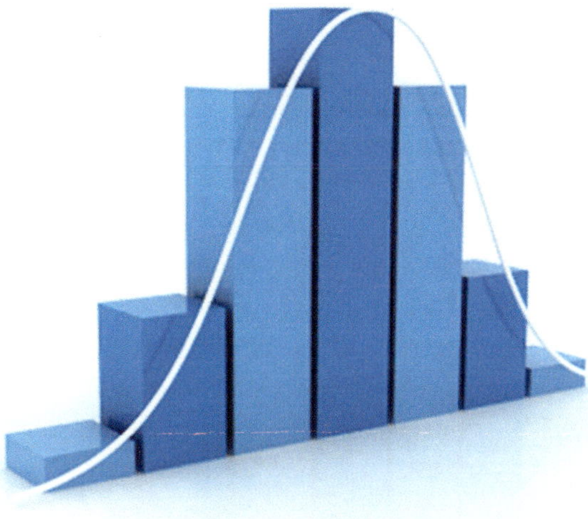

Nearly everyone knows about normal statistical distributions, like the one above.

But skewed statistical distributions, like the one below, are needed to understand how we pay, use, and value electricity.

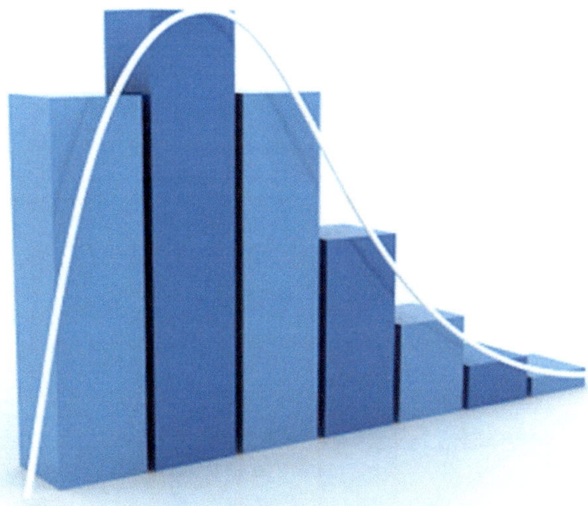

Chapter 3: Prices and Bills

Households' electricity price trends; households' electricity bill trends; low-income households' bill trends

Prove it

Electricity isn't cheap. It's expensive and getting more so all the time.

That's what we hear again and again. Electricity's costs are high and ever-increasing. We hear this from every direction. No one, not even the utilities, says otherwise. It's got to be true.

We can see it in our utility bills. Bills can get to four, five, six hundred dollars or more. This craziness confirms what we hear.

So that's what everybody thinks. Can we convince people to take a second look? Can we convince them that they were wrong? Harder still, can we convince them the truth is the opposite of what they thought?

That's quite a challenge. It won't be easy to turn this situation around. It can't be done overnight. If we're going to convince people electricity's costs are actually moderate and stable, we'd better come up with something really good.

The proof better be compelling. It better be easy to understand. And it better be foolproof. That way, critics will be disarmed, not able to rebut or just muddy the waters.

Rates decreased in thirty-nine states

The US Department of Labor puts out the Consumer Price Index. Everyone calls it the CPI. From 1990 to 2011, over the course of twenty-one years, the CPI increased by 72 percent. This means the cost of living in the country increased by a little over two-thirds.

To keep pace with inflation, your income needed to increase by the same 72 percent. Suppose your income was $30,000 in 1990. Your income needed to increase to $51,600. You'd need a $21,600 income hike over the two decades plus.

What happened to households' rates for grid electricity in this time period? In nine of the forty-eight contiguous states, average rates increased by as much as the CPI or by more. That's right. Rate increases outpaced inflation in just nine states. Rate increases were outpaced by inflation in the other thirty-nine states.[24]

What happened to rates in the remaining thirty-nine contiguous states? Households in these states saw their utility rates lose the race with inflation. Stop the presses! If you live in one of these states, your rates decreased, appropriately adjusting for inflation.

24 Of the forty-eight contiguous states, households' average electricity price per kilowatt-hour fell by over one-third in twelve states during the twenty-one-year period of 1990–2011, after appropriately adjusting for inflation. Households in Arkansas, Arizona, Illinois, Louisiana, New Mexico, and Utah saw huge price decreases. Real prices rose in only nine of the contiguous states, the most populous being Washington and Tennessee, states in which federal government-owned utilities dominate. American households nationally, including Alaska, Hawaii, and the District of Columbia, saw nominal prices rise 50 percent on average. But the Consumer Price Index went up by much more during the period, 72 percent. Real prices fell in every one of the ten most populous states. Among the ten most populous, there were large price drops in California, Florida, Georgia, Illinois, North Carolina, Ohio, and Pennsylvania. Source: Energy Information Administration, Bureau of Labor Statistics.

Graph 1:

Even in California, electricity prices dropped since 1990 (inflation adjusted).

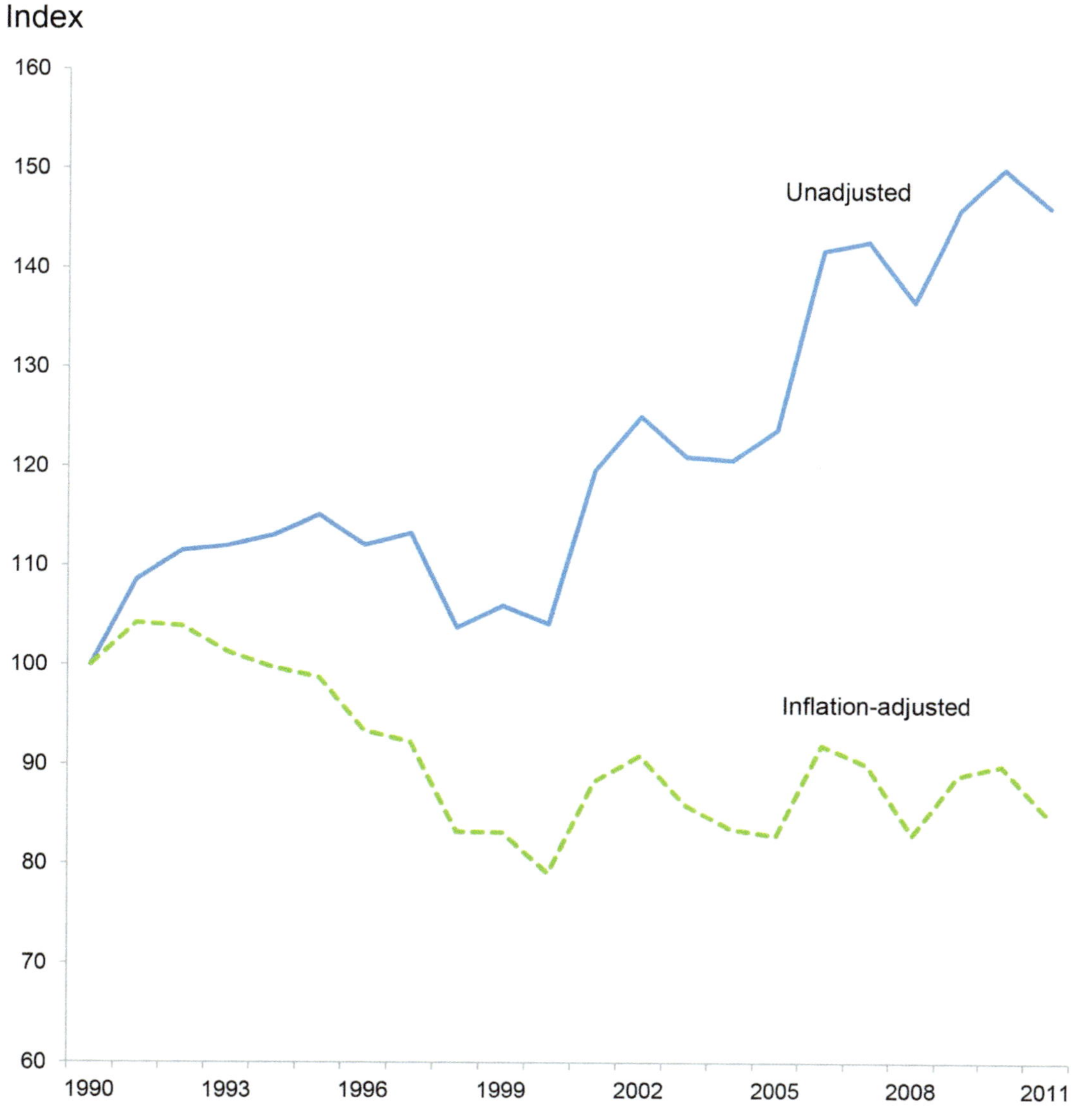

Let's not skip through this point too fast. If you live in one of the thirty-nine states, as do most Americans, the price you pay for electricity has been going down. It hasn't been going up.

You know prices don't go down for that many goods and services. But prices have fallen for electricity. Save the applause for later.

How does this jibe with what you hear from every direction about electricity? Are you still so sure that electricity is expensive? Are you still so sure that electricity is getting more expensive all the time?

Maybe this line about electricity is true for the unfortunate ones in those nine states, maybe. But it's certainly not true for you.

Rates increased in nine states, one really

What about the unfortunate ones? Okay, in those nine states, residential rates increased as much as the CPI did or by more. So, it's true that rates have gone up there, inflation-adjusted.

Of the nine states, rates increased slightly in five of them: Connecticut, Maryland, Montana, Tennessee, and Vermont. Rates in these states increased in the range of 75 percent to 84 percent during the 1990–2011 period while the CPI increased 72 percent. If you live in one of these states, your rates were nearly constant, appropriately adjusting for inflation.

This leaves just four of the forty-eight contiguous states that saw significant rate increases during the last twenty-one years. Which states are they?

Two of the four are in the Pacific Northwest: Oregon and Washington. But rates in the two states were among the lowest nationally in 1990.

Elsewhere there were just two states with significant rate increases. They were Nevada and Wisconsin. Like Oregon and Washington, rates in Nevada were among the lowest nationally in 1990.

That leaves only Wisconsin. Of the forty-eight contiguous states, Wisconsin is the only one in which residential rates significantly increased and rates weren't among the lowest nationally in 1990. So the rate increase story in this one state isn't that great. If you live in the Badger State, at least you have the Packers.

Good times

Now this is remarkable. Rates in twenty-two states increased in the range of 12 percent to 50 percent during the 1990–2011 period, while the CPI increased 72 percent. If you live in one of these states, your rates fell through the floor, appropriately adjusting for inflation.

That's crazy. This has happened to virtually no goods and services in our economy. If you live in one of these states, electricity's cost has been dropping like a rock.

Do you live in Arkansas, Arizona, Illinois, Louisiana, or New Mexico? Did you realize how incredibly fortunate you've been? In these five states, rates increased in the range of 12 percent to 23 percent during the period while the CPI increased 72 percent. Your electricity's cost has been dropping faster than a rock defying the laws of gravity.

The most populous states were invited to the party too. Rates in California, Florida, and Georgia, for example, each increased 48 percent during the period, well short of the CPI's 72 percent.

How low can you go?

Leave out, for a moment, the eleven states in the Northeast and Mid-Atlantic. Add in California to make it twelve states on the list. Of the remaining thirty-six contiguous states, thirty-three of them had average residential rates for grid electricity below the national average in 2011, 11.72 cents per kilowatt-hour.

Who were the three odd ones? There was Illinois, but it hardly counts as an outlier. Illinois rates were just half a percent over the national average, at 11.78 cents. Michigan and Wisconsin were 13 percent and 11 percent over the national average.

Other than them, thirty-three of the forty-eight contiguous states had below-average rates, below 11.72 cents. Their average was around ten cents. Among them were the populous states of Arizona, Florida, Texas, and Ohio.

So if you live in the continental US, outside of the Northeast, Mid-Atlantic, and California, your rate is probably below the national average. It's around ten cents for each kilowatt-hour your household takes.

Think about that. It's awesome. Give me a dime and I'll give you a kilowatt-hour of my finest grid electricity. That's enough juice to light a hundred-watt incandescent bulb for ten hours. Per hour, your cost is a penny.

Rates were extraordinarily economical in fifteen of the thirty-three bargain basement states. In the fifteen states, like Missouri and Washington, residential customers paid below ten cents per kilowatt-hour on average. Consider that for a moment. Like in the old dance song, the question is, how low can you go?

The comparison of states is even more remarkable when you look at grid electricity rates for industrial businesses. Industrial rates are real important because manufacturers sometimes pick up and move to someplace else if they find electricity rates are too high for their liking. In 2011, the average industrial rate was in the narrow range of 5.1–6.7 cents per kilowatt-hour in a large majority of the forty-eight contiguous states, thirty of them. These thirty states were highly competitive for manufacturers, their jobs, and the economic multiplier effects.

As with residential rates, the eleven states in the Northeast and Mid-Atlantic, plus California, Michigan, and Wisconsin, were unhappy exceptions. Their industrial rates were above that narrow range of 5.1–6.7 cents. Three more states were above the range, above 6.7 cents, Colorado, Florida, and Tennessee. One state, Washington, was actually below the range, below 5.1 cents. With its bounty of cheap hydroelectric power plants, it joins the list of thirty-one highly competitive states.

Now on to bills

On a daily basis, a monthly utility bill of $64 is equivalent to $2.13 per day.[25] Simply dividing the monthly bill by thirty for the number of days in a month arrives at this daily amount.

Breaking down a monthly utility bill into a daily average is particularly revealing. The daily average of $2.13 turns out to be an astonishing bargain when measured up against other common daily purchases by consumers.

25 The 2 million households in Colorado paid just $2.12 daily for grid electricity in 2009 on average. The 12 million households in California paid just $2.77. Source: Energy Information Administration, Residential Energy Consumption Survey, 2009.

Graph 2:

Households' electricity expenditures almost unchanged since 1984 despite greater use (inflation adjusted)

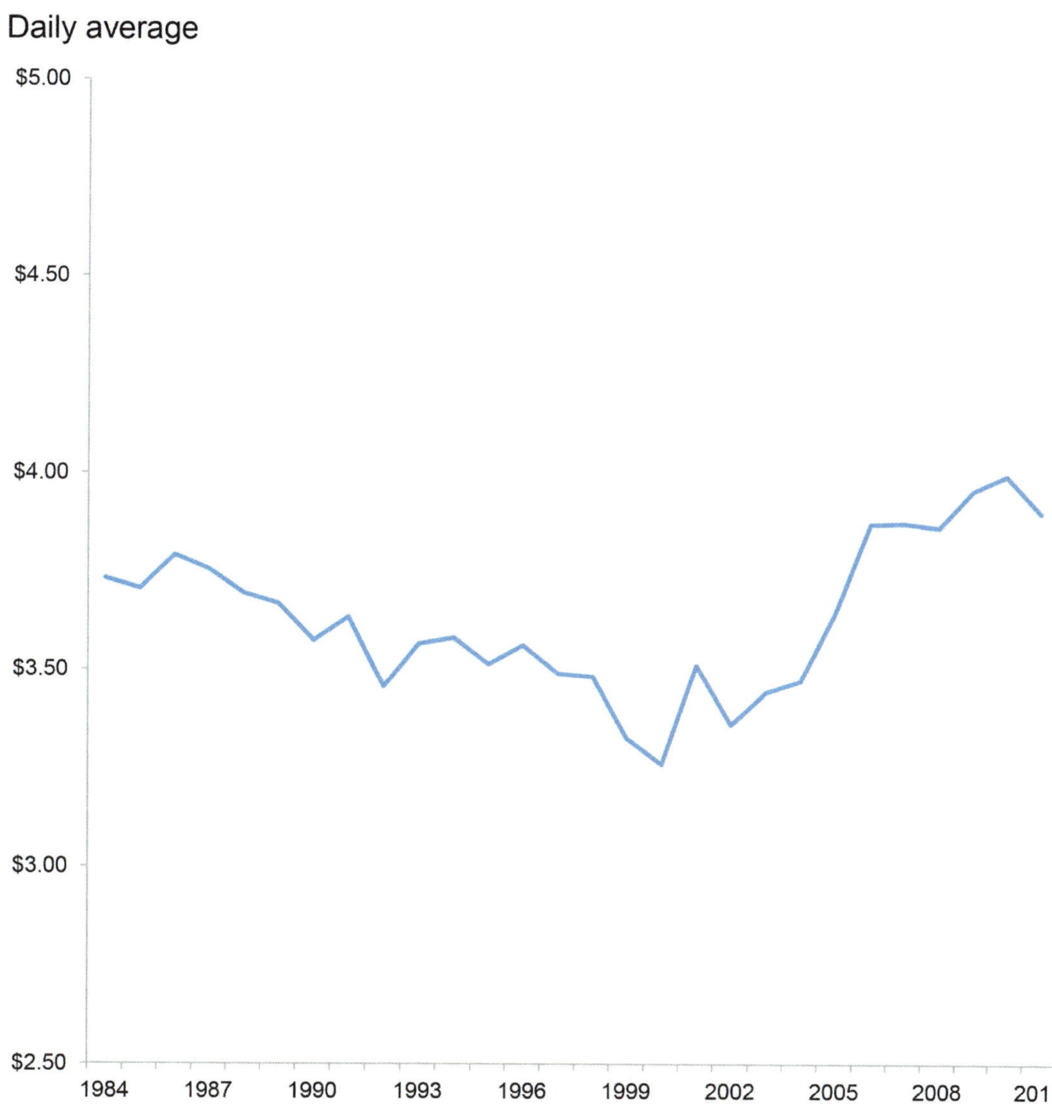

The daily average of $2.13 for all of a home's electricity is roughly equivalent to: the price of a Starbucks venti cup of coffee, or a half-gallon of motor gasoline, or a copy of the *New York Times* at a newsstand (Mondays to Saturdays), or slightly more than a bagel with cream cheese at Dunkin' Donuts.

A consumer could be spending about the same on all of the home's electricity and a daily cup of coffee, or the gasoline to drive the car ten to fifteen miles daily, or a copy of the newspaper, or slightly more than the daily breakfast bagel. It is more incredible when one realizes that this $2.13 daily keeps all of the televisions, video recorders, air conditioners, fans, refrigerators, computers, Internet devices, lights, washers, dryers, microwave ovens, toasters, cell phones, video games, music players, etc. in a home operating twenty-four hours a day and seven days a week.

Bills decreased too

From 1984 to 2011, over the course of twenty-seven years, the CPI increased by 116 percent. This means the cost of living in the country more than doubled since back in 1984.

The Bureau of Labor Statistics goes back that far in its survey of household expenditures, including for grid electricity. So we'll go back that far.

What happened to households' bills for grid electricity in this time period? For all households nationally, their bills increased by 126 percent during the last twenty-seven years. Remember that the CPI increased by 116 percent during the same years. Therefore, bill increases have slightly outpaced inflation since 1984.

The difference between electricity bill increases and CPI increases was indeed slight. The real increase in grid electricity bills per annum has averaged around a third of a percent. Basically, bills have remained virtually constant.

During the last twenty-seven years, households have dramatically increased the number of electrical machines, appliances, and devices in their homes. Households have similarly increased their usage. More use of the grid, at roughly the same cost. That's a pretty good deal for consumers.

Vulnerable

A constant of almost every state public utility commission proceeding deciding about utility proposals to increase rates is the burden the increase would put on low-income households. Much of the media publicity about utility rate proposals highlights how the unemployed, working poor, and lower-middle class could be impacted, as do many of the arguments of utility opponent groups in regulatory proceedings.

Utility regulation staff and commissioners themselves have countless times stated that the impact on these vulnerable households is perhaps their number-one concern in these proceedings. Many regulator decisions have rejected utility rate proposals in whole or in part exactly because this section of the population under financial strain cannot be asked to bear the proposed increase. This has been stated in numerous commission orders.

It is therefore appropriate to focus on the electric rates and bills of the unemployed, working poor, and lower-middle class to the extent that their rates and bills significantly differ from those of more economically fortunate members of society. Have the grid electricity rates and bills of vulnerable households, not just households generally, been high or have they been moderate? And, when utilities propose rate increases, driven by higher operational costs and needs to construct and install equipment, are the impacts on vulnerable households high or moderate?

Graph 3:

Low-income households spend less on electricity; high-income households spend more.

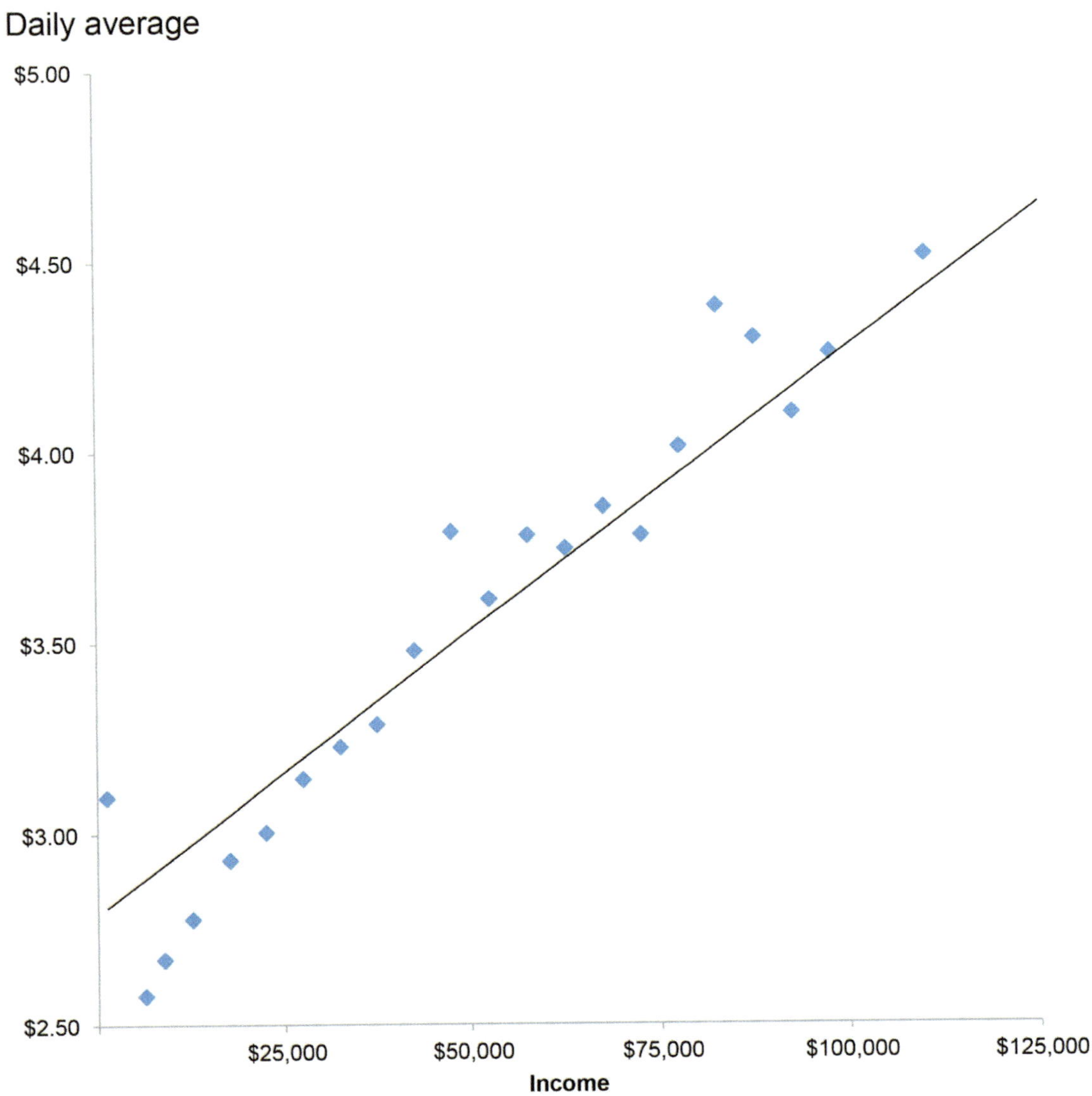

Graph 4:

Most low-income households spend $1.50–$3.50 daily on electricity in Midwest; most high-income households spend more.

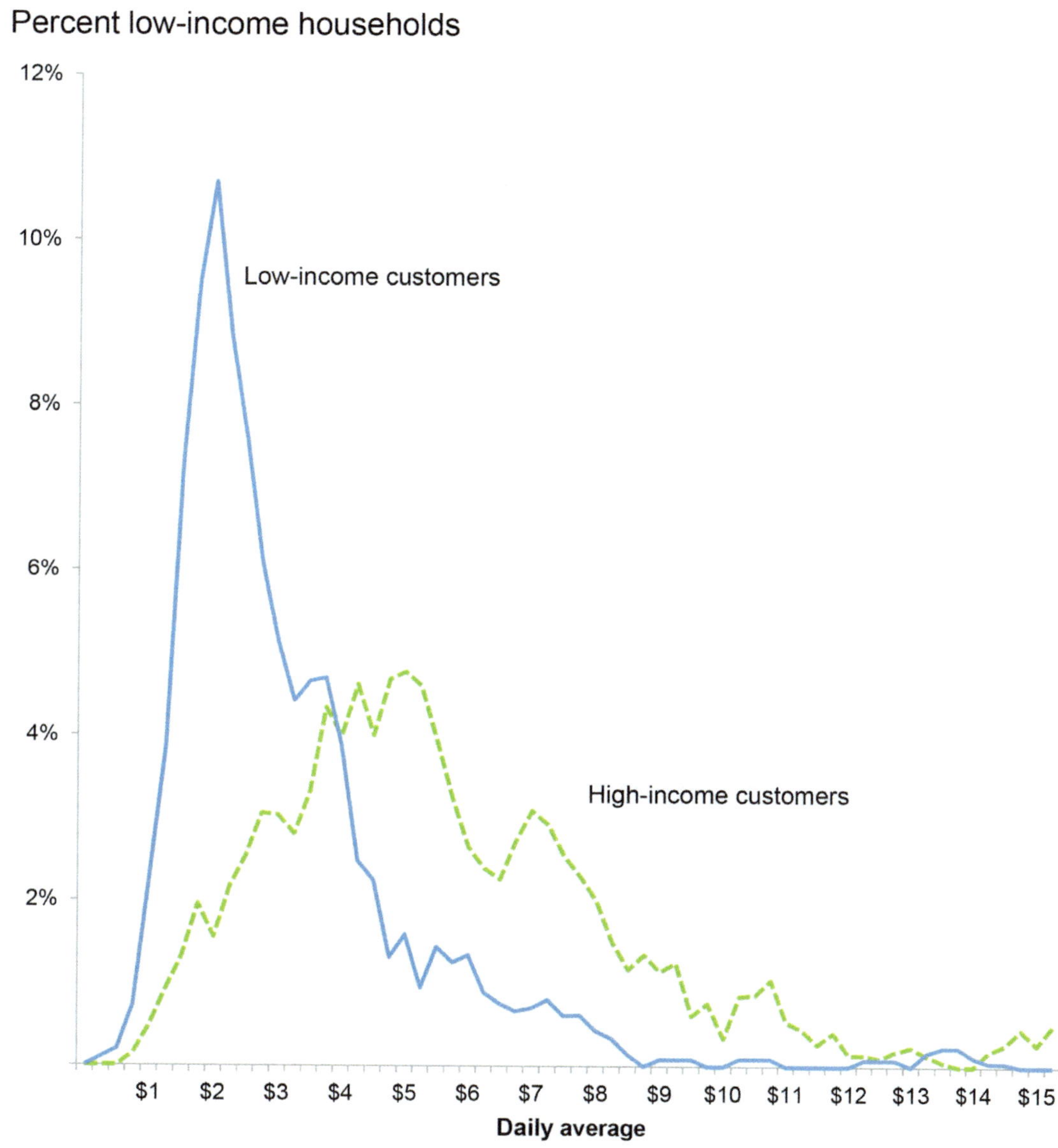

Graph 5:

Households below poverty level in many states outside the South spend under three dollars daily on electricity.

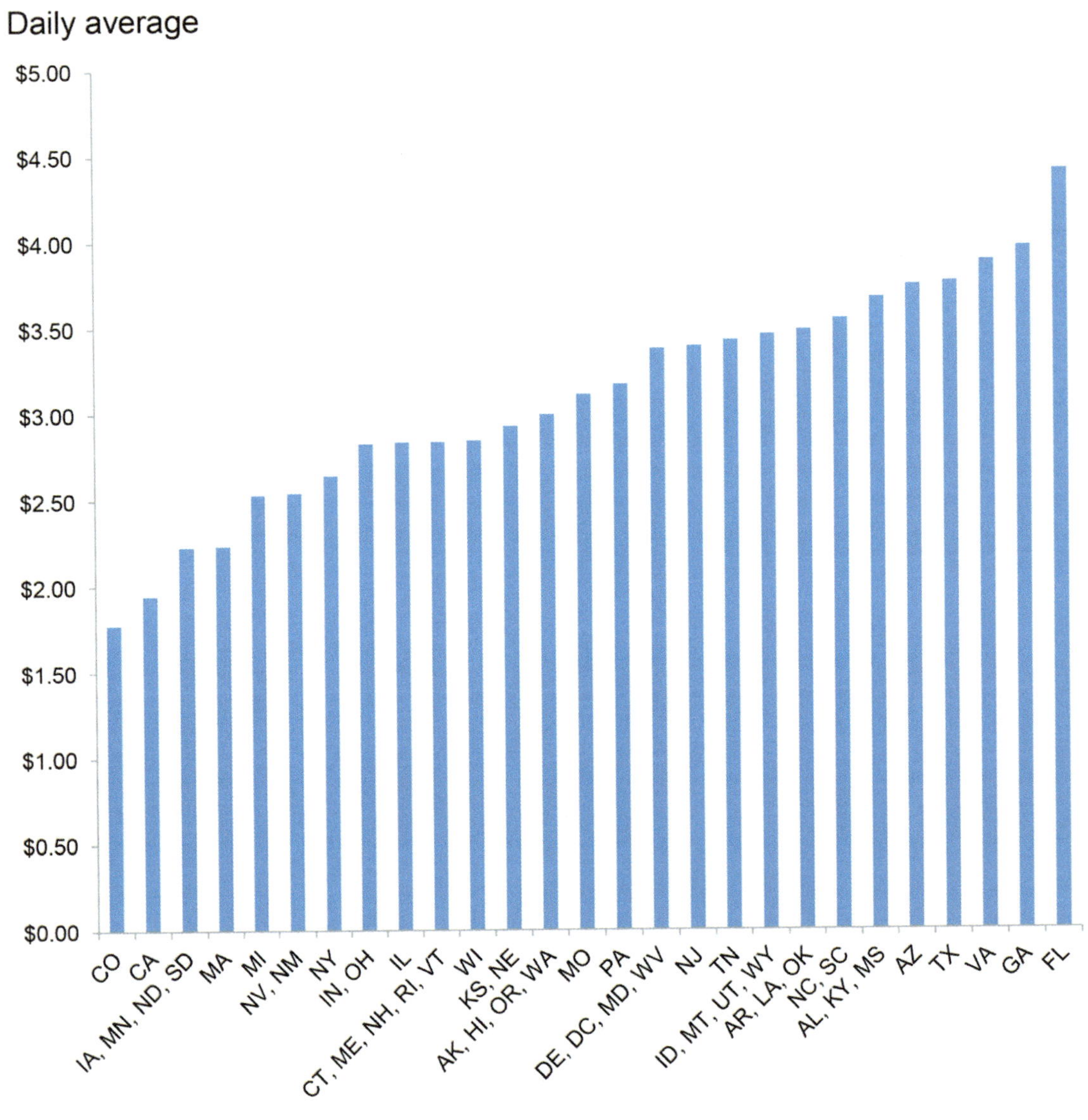

Lowest-income households

Take a look at the lowest-income households. Their incomes are no greater than twenty thousand dollars. There is separate data on these households for sixteen states.

The lowest-income households pay the least for grid electricity in Colorado and California. Their cost is $1.41 daily and $1.80 daily, on average, respectively. These are the most vulnerable households in these two states. Yet $1.41 or $1.80 daily is a modest level of cost for a service as valuable as grid electricity.

It helps to make a comparison to total household expenditures. Total expenditures of these households are around sixty dollars daily, on average. They rely heavily on transfer payments to supplement their incomes.

In contrast to Colorado and California, the lowest-income households in Florida and Georgia pay the most for grid electricity. Their cost is $3.89 and $3.81 daily, on average, respectively. This is more than double. Air conditioning is used more intensively in Florida and Georgia.

The lowest-income households in Texas, Arizona, Virginia, and Tennessee, all Southern states, pay $3.27 daily or over for grid electricity. The lowest-income households in Michigan, Massachusetts, New York and Wisconsin, all Northern states, pay $2.52 daily or under.

So, outside the South, the lowest-income households generally pay less than $2.50 daily for grid electricity. This level tends to be around 2 percent to 3 percent of their total household expenditures.

In the South, where air conditioning is used more intensively, the lowest-income households generally pay more than $3.25 daily. This level tends to be around 3 percent to 4.5 percent of total household expenditures.

Outside temperature determines your electric bill

Electricity use varies quite substantially based on what part of the country a household is located in and also based on the kind of home.[26] Households in the warmer Southern parts of

26 The 25 million households in the West paid $3.01 daily for grid electricity in 2009 on average. In contrast, the 42 million households in the South paid $4.50 daily, 50 percent more. Western households have significantly lower utility bills than Southern households. Source: Energy Information Administration, Residential Energy Consumption Survey, 2009.

the country use much more electricity than those in the cooler Northern parts of the country, whatever their income. Single-family detached houses and mobile homes use much more electricity than households in apartment buildings and single-family attached houses.

There are 15 million households in Florida and Texas, two populous warm-climate states. Florida and Texas households take, on average, 15,862 and 15,149 kilowatt-hours of grid electricity respectively. Their usage per home is fairly high.

Now contrast that with California and New York, two populous temperate-climate states. There are 19 million households in these two states. California and New York households take, on average, 6,992 and 6,882 kilowatt-hours of grid electricity respectively. Their usage per home is fairly low.

So, Florida and Texas households used more than twice as much per home as California and New York households. There's a huge difference.

House type and size determines your electric bill too

Households in single-family detached houses and mobile homes take, on average, 13,159 kilowatt-hours and 11,787 kilowatt-hours. Households in apartment buildings and single-family attached houses take, on average, 7,147 kilowatt-hours and 9,240 kilowatt-hours. So the 79 million households in single-family detached houses and mobile homes used substantially more electricity per household than the 25 million households in apartment buildings and single-family attached houses.

Households that reside in apartment buildings, especially those with five or more units, especially those located in temperate states like New York and California, average well under six thousand kilowatt-hours per annum, well under five hundred kilowatt-hours per month. This comes out to be around 16.5 kilowatt-hours daily.[27]

At fifteen cents per kilowatt-hour, electricity costs this household around $2.50 daily or seventy-five dollars monthly. On summer days when usage is higher driven by air conditioning, electricity costs significantly more than $2.50. In the winter, spring, and fall, electricity costs less than two dollars daily for such households.

27 The 19 million households in apartment buildings with five or more units paid $2.21 daily for grid electricity in 2009 on average. In contrast, the 72 million households in single-family detached houses paid $4.24, 92 percent more. Apartment building homes have significantly lower utility bills than single-family detached houses. Source: Energy Information Administration, Residential Energy Consumption Survey, 2009.

Bills by state

Residential electricity use in five states (California, Maine, New York, Rhode Island, and Vermont) averaged 521 to 581 kilowatt-hours monthly or 17.4 to 19.4 kilowatt-hours daily. These low usage levels are notably averages for all households in these states, including high-income households in single-family detached houses.

The average residential electric bill for all households rich and poor in four of these five states, all but New York, was $81.60 to $88.83 monthly or $2.72 to $2.96 daily. Again, most of the less economically fortunate paid closer to $2.50 daily.

Nationally, the average residential electric bill for all households in sixteen states was in the eighty- to ninety-dollar range. In six more states, the average bill was actually lower than eighty dollars monthly or $2.67 daily.

Average daily bills averaged three dollars or more in the remaining twenty-eight states and four dollars or more in eleven states: Alabama, Connecticut, Delaware, Florida, Hawaii, Maryland, Mississippi, Nevada, South Carolina, Texas, and Virginia. Because of the distribution of electricity use, in which the unemployed, working poor, and lower-middle class generally consume less than the average, we can figure that many of these households pay in the range of three dollars daily.

Households with incomes of less than ten thousand dollars, for instance, take only half as much grid electricity as do those with incomes of $100,000 and up, approximately. Those households with incomes of less than ten thousand dollars take only two-thirds as much as does the average American household, again approximately. For two households in the same community with the same electric rates, the low-income household could be paying half as much daily and monthly for electricity as the wealthy household.

Two dollars and fifty cents too much?

No matter how we measure it, except in the South, most low-income households pay around $2.50 per day on average for all of their electricity twenty-four/seven.[28] This amount, $2.50 per day, is not zero. Grid electricity isn't free for low-income households. However, one

28 The 24 million households with income less than $20,000 paid $2.85 daily for grid electricity in 2009 on average. In contrast, the 12 million households with income greater than $120,000 paid $5.43, 91 percent more. Low-income households have significantly lower utility bills than high-income households. Source: Energy Information Administration, Residential Energy Consumption Survey, 2009.

could certainly say that $2.50 per day is generally not burdensome even for low-income households. And one could certainly say that $2.50 per day is proportionate for what low-income households receive from their utility and the grid.

Should utility regulators strive to make grid electricity even less expensive for low-income households? Should regulators drive the $2.50 per day to $2.25 per day? Some might advocate this. But what would be sacrificed to save low-income households twenty-five cents daily? Would be the ability of the grid to deliver uninterrupted electricity be compromised in the process?

This result is important. It is the burden of utility bills and proposed rate increases on these low-income households that animates most of the controversy in utility regulatory proceedings. If what low-income households pay for grid electricity is not burdensome and is proportionate, then these facts should factor in the decisions of regulators.

Two Kit Kats, box of Cheerios

Tens of millions of American households spend three dollars daily for the grid electricity they take. Many of them spend $2.50 or less. Is that a lot or a little for the comfort, convenience, and entertainment the grid electricity provides daily?

Why don't we compare this level of expenditure with what else it buys for us? For $2.50, you can buy two Kit Kat bars (my favorite before my diet kicked in). There are many goods and services we buy frequently that are in the range of $2.50 to $3.00. A single subway ride in New York or Washington, DC, for instance, costs around this much, as does cable TV daily. Do they mean as much to us as our home's power? How about two bottled waters or a gallon of milk on sale? Does their production and delivery to us require as much state-of-the-art technology and skilled personnel?

It's astounding. For around the cost of two Kit Kat bars, tens of millions of households are able to run all of their electrical machines, appliances, and devices daily.

Graph 6:

Households' daily electricity cost low compared with common purchases

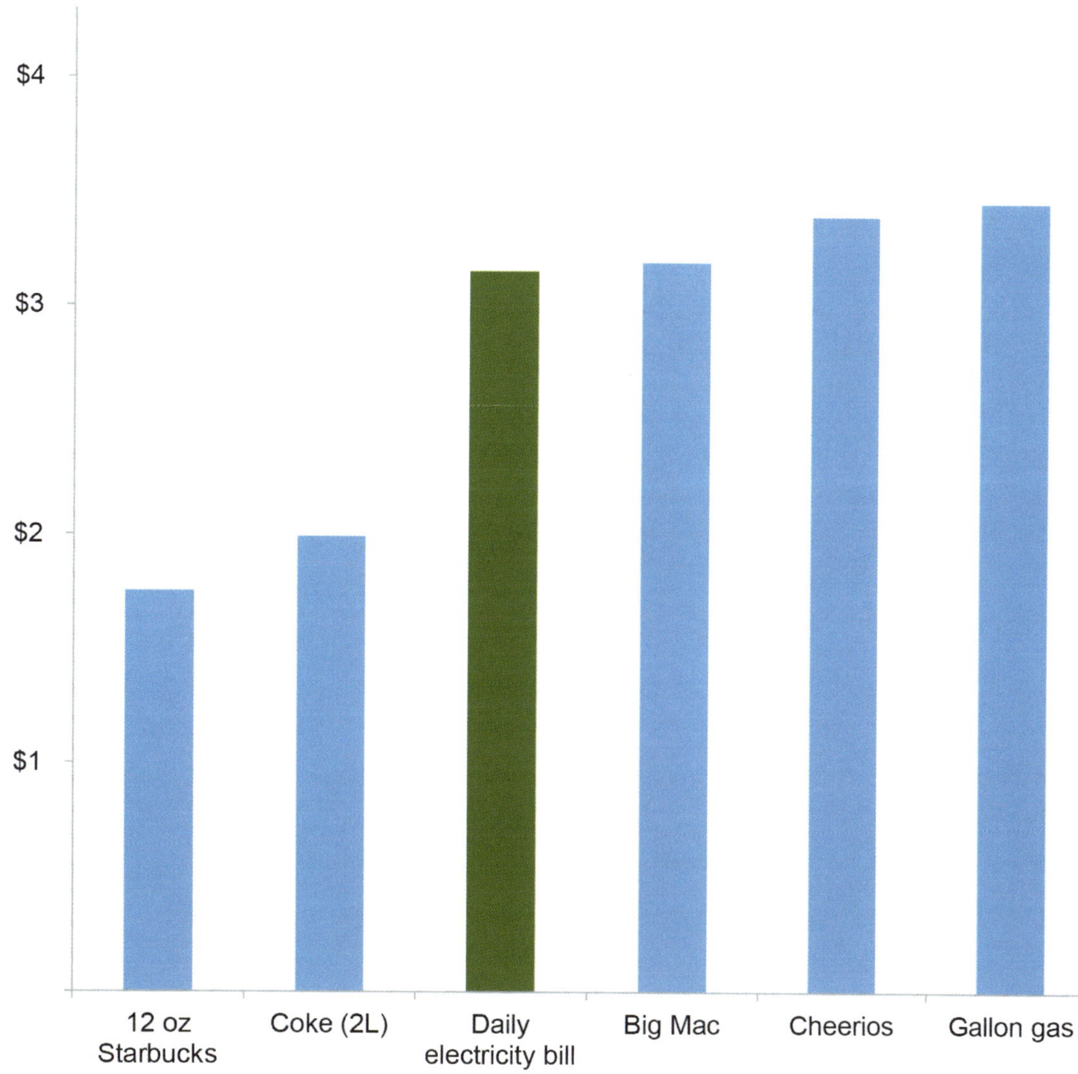

Reddy's Recharges

1. Grid electricity's cost is moderate and stable, and this is easily demonstrated.

2. In most states, household electricity rates have either *decreased* or held steady for decades (appropriately adjusting for inflation).

3. Electricity has consistently been less than 3 percent of Americans' household expenditures and around 4 percent for low-income households.

"Are you just pissing and moaning, or can you verify what you're saying with data?"

Chapter 4: Breaking It Down, Daily

Households' grid electricity bills generally two to four dollars daily; bills vary with temperature; averages exaggerate bills

Comparison shopping

There is a broad variation in how much electricity households take from the grid. This is true at any utility in the United States. Thus, there is a broad variation in how much households pay for the grid at any utility.

The broad variation occurs even though all of a utility's households are billed in strict accordance with a single uniform rate structure. One household uses six hundred kilowatt-hours in a month and pays sixty dollars, while a neighbor uses eighteen hundred kilowatt-hours, three times as much, and pays $180, three times as much. Not far away, a third household served by the same utility uses three thousand kilowatt-hours and pays three hundred dollars. The third household takes and pays five times as much as the first. They're members of the same community, they're served by the same utility, yet they vary so much in what they take and pay.

But try comparing utilities. Will we find the same broad variation between utilities? The answer is no.

The US Energy Information Administration tracks what households take and pay at 407 utilities nationally. Let's see how the average households of utilities compare.

Utilities' average households

The average household served by Nevada Power in Nevada paid 11.9 cents per kilowatt-hour of grid electricity in 2011. We cite Nevada Power because its residential customers were the national median with respect to the average per-kilowatt-hour rate.

The median is a handy statistic. It splits data into two groups of equal numbers, two halves. For example, the median batting average in the American League might be .260. Half the hitters have batting averages above .260 and half have averages below .260. If John "Blackout" Barnes had a .260 average, we'd say he's at the median.

The median here splits all American households in two halves. Half of all households paid a higher rate than the median and half paid a lower rate than the median, based on utilities' average households.

The average household served by West Penn Power in Pennsylvania paid 9.8 cents per kilowatt-hour. We cite West Penn Power because its residential customers were at the dividing line for the bottom quintile nationally. Eighty percent of American households paid a higher rate than the bottom quintile, and twenty percent paid a lower rate.

The average household served by West Penn Power paid 2.1 cents per kilowatt-hour less than did the average household served by Nevada Power. Suppose you use twenty kilowatt-hours daily. The difference in rates between West Penn Power and Nevada Power would amount to just forty-two cents daily.

The average household served by Pacific Gas & Electric in California paid 15.5 cents per kilowatt-hour. We cite Pacific Gas & Electric because its residential customers were at the dividing line for the top quintile nationally. Twenty percent of American households paid a higher rate, and eighty percent paid a lower rate than the top quintile.

The average household served by Pacific Gas & Electric paid 3.6 cents per kilowatt-hour more than did the average household served by Nevada Power and 5.7 cents more than the average household served by West Penn Power. Suppose you use twenty kilowatt-hours daily. The difference in rates between Pacific Gas & Electric and Nevada Power would amount to just seventy-two cents daily. The difference in rates between Pacific Gas & Electric and West Penn Power would amount to just $1.14 daily.

It's a significant difference, $1.14 daily, but then we're comparing two utilities at opposite ends of the spectrum. One, West Penn Power, is at the dividing line for the bottom quintile nationally. The other, Pacific Gas & Electric, is at the dividing line for the top quintile nationally.

The average households' per-kilowatt-hour rates and daily costs of most US utilities are tightly clustered around the median—at Nevada Power—and below. To illustrate, the average household served by Sierra Pacific Power, also in Nevada, is at the dividing line for the next-to-the-bottom quintile. The average household there paid 11.1 cents per kilowatt-hour.

The average household served by Commonwealth Edison in Illinois is at the dividing line for the next-to-the-top quintile. The average household there paid 12.6 cents per kilowatt-hour.

So there's only a 2.8 cents spread between West Penn Power, down at the bottom quintile, and Commonwealth Edison, up at the next-to-the-top quintile. Again, suppose you use twenty kilowatt-hours daily. The difference in rates between West Penn Power and Commonwealth Edison would amount to just fifty-six cents daily.

This is a remarkable result. Whether you are the typical household served by a utility with one of the nation's cheapest rates or by a utility with one of the higher rates, the bottom line difference is just fifty-six cents daily. There's extraordinary consistency between US utilities as to what they charge per unit of grid electricity.

Drill, baby, drill

Let's drill down deeper on this. Let's compare the average household's daily cost of electricity at different utilities.

This time, the median was Duke Energy Carolinas in North Carolina. The average household served by Duke Energy Carolinas paid $3.36 daily for grid electricity in 2011.

Which utilities this time were at the dividing lines of the quintiles nationally? Down at the low end were Commonwealth Edison and Pacific Gas & Electric, roughly tied at the dividing line for the bottom quintile. (Yes, we already saw these same utilities in the comparison of per-kilowatt-hour rates.) Their average households paid $2.84 daily. Wisconsin Electric Power in Wisconsin was at the dividing line for the next-to-the-bottom quintile. Its average household paid $3.17 daily.

Above the median, Duke Energy Carolinas in South Carolina and Central Hudson Gas & Electric in New York were roughly tied at the dividing line for the next-to-the-top quintile. Their average households paid around $3.60 daily.

Once again we see the clustering of most US utilities. There's only seventy-six cents daily separating Commonwealth Edison and Pacific Gas & Electric at the low end, at the dividing line for the bottom quintile, from Duke Energy Carolinas and Central Hudson Gas & Electric at the high end, at the dividing line for the next-to-the-top quintile.

Only seventy-six cents daily, yet we're including 40 percent of all American households within this range. Put another way, the average household of the utilities that serve nearly a majority of Americans paid $3.21 daily for grid electricity plus or minus just thirty-seven cents.

How do these facts jibe with the common criticism that a given utility has outrageous rates? The reality is that American utilities generally incur similar costs, within a rather narrow range. Their rates and households' utility bills are accordingly similar, for a level of electricity usage.

Per-kilowatt-hour rates are significantly higher in densely populated regions such as in the Northeast and California. There are various reasons for this tendency, particularly state policies in these regions that drive up utility costs. Nevertheless, because households in densely populated regions take less grid electricity on average, average utility bills are not very different from those in less densely populated regions with lower per-kilowatt-hour rates.

That's why there's so much consistency in average household electricity costs coast to coast. Some utilities have somewhat lower averages and some somewhat higher averages, but generally a utility's average household pays a little bit over three dollars daily for its grid electricity.

What gives?

There are exceptions. Most prominently, there are the regions in which air conditioning is used intensively. This makes average households' take from the grid unusually high.

Cleco Power in Louisiana illustrates the point. The average household served by Cleco Power paid $5.31 daily for grid electricity in 2011. No other investor-owned utility in the contiguous forty-eight states had a higher average.

What gives with Cleco Power? Why was the Cleco bill $1.95 daily more than the median for utilities nationally?

It's not the rates charged by Cleco Power and set by the Louisiana Public Service Commission. The average household served by Cleco Power paid a per-kilowatt-hour rate of 11.8 cents. The median rate for utilities' average households nationally, again, was almost identical at 11.9 cents.

So Cleco Power's rate is nice and medium in comparison to the whole country. Cleco Power's bills are relatively high not because of its rate but because households in the heart of Cajun country are comparatively heavy users of electricity.

The average household served by Cleco Power took 1,367 kilowatt-hours monthly from the grid in 2011. That comes to forty-six kilowatt-hours daily. In contrast, the average household served by Pacific Gas & Electric took 559 kilowatt-hours monthly. That comes to nineteen kilowatt-hours daily. This is a huge difference in electricity usage.

The difference can be even greater. The average household served by Consolidated Edison in New York took 420 kilowatt-hours monthly. That comes to fourteen kilowatt-hours daily. This means that the average household in New York City and its northern suburbs took well less than one-third of the electricity than did the average Cajun household.

Here's another way to look at it. If you could persuade three typical New York City households to cut their electricity use by 10 percent, that would save the grid from producing a little over four kilowatt-hours daily. Instead, if you could persuade one typical central Louisiana household to cut its use by 10 percent, that would save the grid significantly more. If you're into energy efficiency, you should be spending more of your time in Avoyelles, Natchitoches, and Rapides parishes than in the Bronx, Brooklyn, and midtown.

Thirteen cents

Utilities charge you around a hundredth of a penny for a watt-hour. That's if you take a watt from the grid continuously for an hour. They charge you around ten pennies, a dime, for a thousand watt-hours. That's if you take a thousand watts continuously from the grid for an hour.

You could take half a thousand watts for two hours. You could take two thousand watts for a half hour. It's all the same. That much electricity, over that much time, a thousand watts for an hour or the equivalent, is called a kilowatt-hour.

How much electricity is a kilowatt-hour? Is it a lot? Here's a handy way to think about it. One kilowatt-hour is enough to power everything in the average American home for three-quarters of an hour. That's everything in the home.

That one kilowatt-hour air conditions the home if it's summertime. It heats the home if it's wintertime and if the home has electric heating. That same kilowatt-hour also powers all of the lights, computer equipment, televisions, video recorders, video games, audio systems, and mobile device chargers. That same kilowatt-hour also powers the washer/dryer, refrigerator, microwave oven, hair dryer, and toaster. All of the many machines, appliances, and devices you plug in are completely covered for the three-quarters of an hour.

A dime gets you that kilowatt-hour. A kilowatt-hour gets you grid electricity for three-quarters of an hour. On an hourly basis, this comes to about thirteen cents per hour. On a daily basis, this comes to about $3.20 per day. For thirteen cents hourly, or $3.20 daily, the average household takes whatever it wants and needs from the grid.

Half of thirteen cents

Did you know that half or more of the money that you use to pay your utility bill doesn't really go to your local utility? Though the utility collects it when you pay the bill, half or more of the money is immediately turned over to others. The percentage varies significantly by state. The grid's power plant owners get a chunk for the manufacture of kilowatt-hours in bulk. In many states, under power plant deregulation, the grid's power plant owners are separate companies from your utility. State and local governments get a chunk for various taxes and fees for a range of public policy purposes.[29]

The remaining half or less of the money does go to your local utility. That's half of thirteen cents hourly from the average household, or half of $3.20 daily. When you wonder whether your local utility should be doing a lot more, to quell power outages for example, remember that all you're paying the utility is six and a half cents hourly or $1.60 daily. For that amount of money, what do you expect?

29 Steven Mufson, "Cheaper Natural Gas Doesn't Guarantee Lower Fuel Bills for Homes," *Washington Post*, Dec. 15, 2012. "Pepco says 72.2 percent of the typical electricity bill in the District [of Columbia] is made up of fuel costs, while *distribution eats up [only] 19.2 percent*; transmission 2.2 percent; and surcharges and taxes, 6.4 percent." *Emphasis added*

It's not that much money, you have to admit. Imagine if your utility had the temerity to propose to regulators to increase resources for the utility's stewardship of the local grid by a whopping 20 percent. That doesn't seem like chump change, a 20-percent increase. Yet, in terms of how much more households would have to pay for this increase, it would come to 20 percent of six and a half cents hourly. That's 1.3 cents hourly. On a daily basis, that's thirty-one cents. Many households would willingly pay thirty-one cents more per day if the extra amount would really ensure there would be fewer and shorter outages.

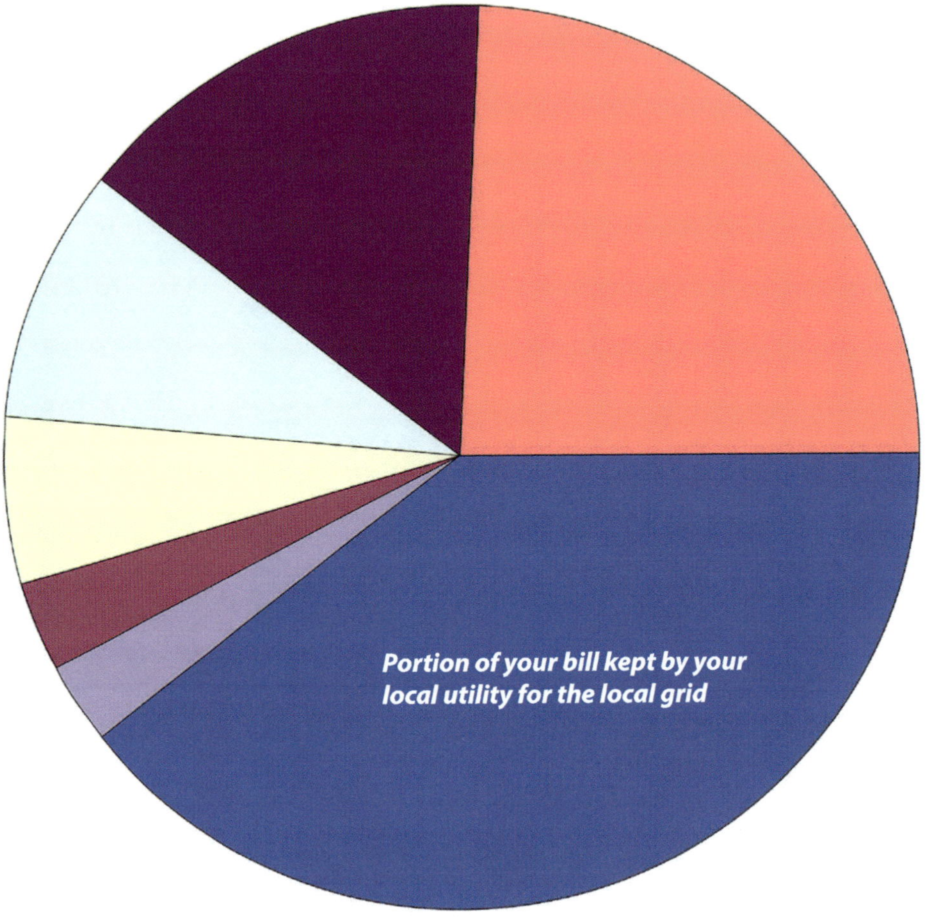

Did you know that half or more of the money you use to pay your utility bill doesn't really go to your local utility? Though the utility collects it when you pay the bill, half or more of the money is immediately turned over to others.

Mean average

Let's look at how households' utility bills vary within the area served by a single utility. The statistical distribution of what households pay for grid electricity can be illustrated with a simplistic example.

Suppose an impossibly small utility has just four residential customers. One customer is a light user of electricity. One customer is a heavy user of electricity. The other two customers are medium users.

The light-usage household pays two dollars per day for its grid electricity. It takes just sixteen kilowatt-hours per day from the grid.

The two medium-usage households pay $2.50 per day for their grid electricity. They take twenty kilowatt-hours per day from the grid.

The heavy-usage household pays five dollars per day for its grid electricity. It takes forty kilowatt-hours per day from the grid.

Our impossibly small utility gets paid a total of twelve dollars per day to operate and maintain its corner of America's grid. The utility supplies ninety-six kilowatt-hours per day to its four customers. Each customer pays the uniform per-kilowatt-hour rate of 12.5 cents per kilowatt-hour.

The median daily cost of electricity is $2.50. The number of customers with cost of electricity above $2.50 daily is equal to the number with cost below $2.50.

However, the mean or average daily cost of electricity is three dollars. This is true even though three-quarters of the customers pay $2.50 or less. It's true even though one-quarter of the customers pay just two dollars.

It's true because one customer's average daily cost of electricity and electricity usage is well above the mean. It's higher by a whole 67 percent. The one customer is only one-quarter of all of the customers. But, all by itself, it drives up the mean well above the median.

Is this skewing of residential bill distributions, toward the heavy-user tail of the distributions, characteristic of all utilities? Yes, this is a fundamental characteristic of all utility's residential bill distributions. This is a crucial point in electricitynomics.

Heavy-usage households tend to use much more electricity than the mean usage level. Light-usage households, on the other hand, tend to use only a little bit less electricity than the mean usage level. The implications are huge.

Normal world

The heavy-usage household averages a daily cost of electricity of five dollars per day. That's 67 percent above the mean for all four customers, three dollars per day. In other words, the heavy user takes well more electricity from the grid than the mean and thus pays well more than the mean.

It would be difficult to imagine an offsetting light-usage household. Can a light user have an average daily bill that is 67 percent below the mean, three dollars per day, for all four customers? It wouldn't be that easy.

That offsetting light-usage household's average daily costs would need to be just one dollar per day. It would need to take just eight kilowatt-hours per day from the grid. Then the household paying one dollar per day and using eight kilowatt-hours per day would nicely offset the household paying five dollars per day and using forty kilowatt-hours per day.

The two households would exactly balance. In the center would be the mean of three dollars per day and twenty-four kilowatt-hours per day. The low-end left-hand tail and high-end right-hand tail of the distribution would be perfectly balanced. As a graph, it would be a classic bell curve. Such a symmetric distribution is called "normal."

Abnormal world

A few American households average daily costs for grid electricity as low as one dollar per day. It would mean they take just eight or so kilowatt-hours per day from the grid.

Eight kilowatt-hours daily comes out to an average of 333 watts hourly. That's a dribble of power. You could have three light bulbs on, but that would be about it.

So the low-end left-hand tail of the distribution of utility bills and electricity usage is not that densely populated at one dollar and eight kilowatt-hours daily. Few households are over there. It's nothing like the density at the high-end right-hand tail of the distribution. Plenty of households are over there.

A fair number of American households average daily costs for grid electricity as high as five dollars per day. They take as much as forty or so kilowatt-hours per day from the grid.

Indeed, a number of households average even higher daily costs and electricity usage. They take much more from the grid, and they pay for it.

So the high-end right-hand tail of the distribution of utility bills and electricity usage is densely populated at five dollars and forty kilowatt-hours daily. And there are even more extremes of daily costs and electricity usage out there on the right. As a consequence, the high-end right-hand tail of the distribution easily out-balances the low-end left-hand tail.

Seasonal variation

We said the median average household, among all 407 utilities nationally, paid $3.36 daily for grid electricity in 2011. Multiply $3.36 by thirty for the approximate number of days in a month. The multiplication yields $100.80. So $3.36 daily translates into a monthly utility bill of around $100.80.

Did the median average household receive identical utility bills for all twelve months of 2011? No way, you know it didn't. Some of the monthly bills were higher than $100.80, and the rest were lower. It would be a statistical aberration if even one of the monthly bills were exactly $100.80.

Most of the monthly bills were lower than $100.80. There weren't six monthly bills higher and six lower. That's because a couple of the bills during summer months were way higher than $100.80. They might have exceeded $200.

That's all it takes to push up the annual average to $100.80. It's the way the arithmetic works. Assume that a household gets really high utility bills in July and August, reflecting its intensive air conditioning use in summertime. Assume each of these bills were $300. The daily cost amounts to $10.

The implications are incredible. Half the household's annual electricity costs were incurred in those two months. The other half were incurred in the remaining ten months, spread over them.

With the July bill being $300 and the August bill being $300, the average bill during the remaining ten months was just $60.96. The daily cost amounts to $2.03, a little more than one-fifth of the summertime daily cost.

Graph 7:

Most households' monthly utility bills are low, but most people remember best the few high summer bills.

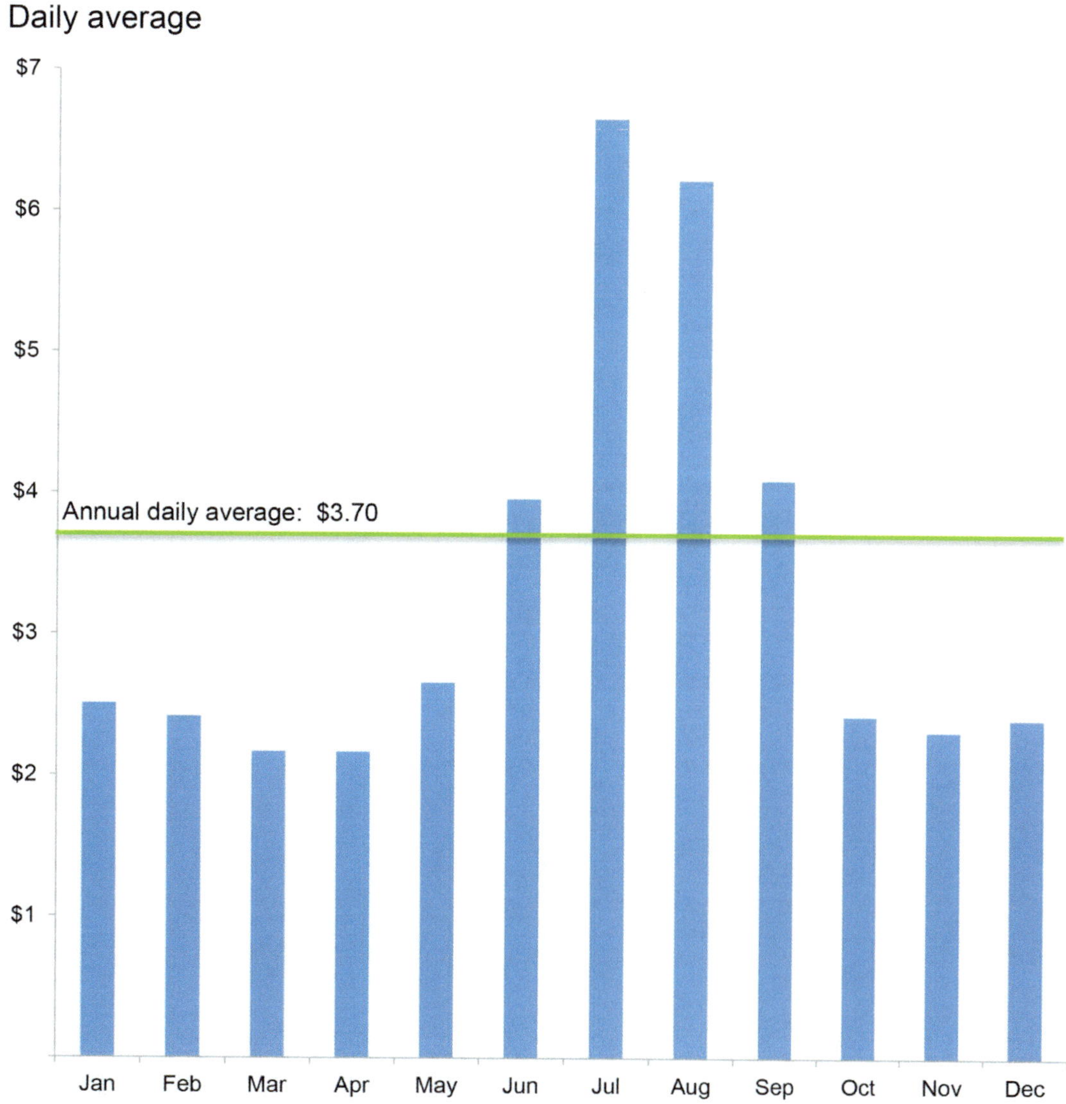

Most people don't realize this seasonal dynamic. People not realizing seasonality drives utilities crazy.

Imagine if more people did get it. These folks would better understand that that their two really high summertime bills were unrepresentative of their electricity costs throughout the year. And they would better understand that the real force behind utility bills is the temperature outside and, well, their own preferences for air conditioning.

Household variation

The light-usage household in the earlier simplistic example pays an average of two dollars per day for grid electricity. It takes an average of sixteen kilowatt-hours per day from the grid. That's not very much electricity to run a household that lives in a single-family detached house. Actually, it's real uncharacteristic.

But sixteen kilowatt-hours per day is more than enough to power most households that live in multi-family buildings. Chances are our light-usage household lives in a multi-family building.

While the daily average over all the days of the year is two dollars, the actual amount varies wildly from day to day and month to month. This extreme fluctuation in grid electricity usage and cost is a crucial point in electricitynomics. The volatility greatly impacts what people perceive they use and pay.

In June, July, and August, three summertime months, our light-usage household pays an average of $4.67 per day for grid electricity. That's well above the average for the whole year, two dollars. Indeed, it's more than double the average for the whole year.

Hourly, it's nineteen cents. For all of the hours of these three summertime months, our light user pays nineteen cents per hour for its electricity.

In contrast, in February, March, April, October, and November, five spring and fall months, our light-usage household pays an average of eighty-three cents per day for grid electricity. That's well below the average for the whole year, two dollars. Indeed, it's less than half the average for the whole year.

Hourly, it's three cents. For all of the hours of these five spring and fall months, our light user pays three cents per hour for its electricity.

In the three summertime months, it's nineteen cents per hour. In the five spring and fall months, it's three cents per hour. That's quite a large difference. Our light-usage household pays over six times more for its electricity in the three summertime months than in the five spring and fall months. Or, put another way, our household pays less than one-sixth in the spring and fall than in the summertime.

It depends on how you look at it. Many Americans decry how high their summertime utility bills are. Few Americans savor how low their spring and fall utility bills are.

Fill it up

During five of the twelve months of the year, our light-usage household pays eighty-three cents daily for all its electricity. Hardly anything else our household purchases on an everyday basis costs as little as eighty-three cents. Think about it.

It can buy about a fifth of a gallon of motor gasoline for eighty-three cents. That's less than a quart of gas, only enough to drive five miles or so. How is this even possible? How can electricity be so cheap that on a daily basis it costs less than a quart of gas?

Or compare a household's expenditures on electricity and motor gasoline another way. Look what happens when it buys a tank of gas. That single fill-up is roughly equal to what the household pays for two spring or fall months of electricity!

Rate un-shock

Suppose utility regulators decided to increase the per-kilowatt rates a utility charges our light-usage household. Suppose further the increase is a whopper, a 10-percent hike. Many incensed critics would label this outrage a "rate shock."

The term "rate shock" implies the 10-percent hike would really jolt households like our light user. But during the five spring and fall months, the hike would have our light user pay around $2.50 more monthly for all its electricity. That comes out to around eight cents more daily. Per day, it would have to pay ninety-one cents, up from eighty-three cents. It's an increase, to be sure. But not the sting the critics made it out to be.

The main impact of rate hikes on households is to raise the costs of air conditioning their homes during summertime months. This is another crucial point in electricitynomics.

What's really happening when utility regulators hike utility rates? Where's the real effect? Few if any regulators realize the bull's-eye in utility rate cases is what it costs to air condition away the summer heat in people's interiors. When regulators hike rates, they're primarily making that a little more expensive.

The rate hike's effect on the other ways households use electricity is fairly small change. The rate hike's effect on households outside of the summertime months is small change too.

During the three summertime months, our light-usage household would have to pay around fourteen dollars more per month for its electricity. That comes out to around forty-seven cents more per day.

The so-called rate shock raises electricity costs eight cents daily during the five months when air conditioners are left off. It raises electricity costs forty-seven cents daily during the three months when air conditioners are run hard. What a difference!

With the exception of air conditioning, households like our light user pay little for every other use of electricity singly or in total. So when regulators raise rates, what these households pay for every non-air-conditioning use of electricity is little affected.

Reddy's Recharges

1. Grid electricity's daily cost provides perspective about electricity's affordability.

2. Electricity's cost varies broadly among the customers of any utility, and it varies broadly from season to season.

3. Utility rate increases mainly impact households' cost to air condition during the summertime.

"Frankly, Harold, you're beginning to bore everyone with your statistics."

Chapter 5: Broad Variation in Bills and Use

Large majority of households pay less-than-average bills; skew from minority of heavy takers; light takers benefit

All the children are above average

Up to now, we've generally been looking at a utility's households as a whole. Our metrics of how much electricity a utility's households take from the grid and how much they pay for what they take are metrics of the whole.

The average is the most common metric of a whole distribution. It usually captures reasonably well what a whole group of anything is or does. The average would just as well capture what a whole group of households served by a utility is or does if...

For the average to be an indicative metric, a whole group must have certain basic characteristics. In statistical terms, a whole group must be distributed normally.

In a normal distribution, most group members fall in the middle. What they are or do is medium, middle of the road. Equal numbers of the group's members fall to the right and to the left. What they are or do is higher than the medium (for the members falling to the right) or lower than the medium (for the members falling to the left). The higher-than-mediums and lower-than-mediums exactly mirror and balance off each other. A normal distribution's symmetry is exquisite.

Guess what? That's not how a utility's households are distributed. The distribution isn't normal if we're looking at how much electricity households take from the grid. It isn't normal if we're looking at how much they pay for what they take.

We'll define some terms to make this clearer. Some households in any utility's area are "heavy users" of grid electricity. They take significantly more than the average for that utility. A number of these heavy users are the "heavier users." And a number of these heavier users are the "heaviest users." Think of them as the sumo wrestlers of electricity usage.

The quantities of electricity that heavier-usage households and heaviest-usage households take from the grid are remarkable. Heavier users take two to three times the average taken by all of the households. Heaviest users take three to four times the average taken by all of the households, or even more. The utility lines to their homes carry plenty of power.

Some households in any utility's area are "light users" of grid electricity. Light users take significantly less than the average for that utility.

Unlike the heavy users, light users don't have extreme subsets among them in noteworthy numbers. There is no light-user equivalent to heavier users and heaviest users. The terms "lighter users" and "lightest users" don't make much sense. Some households take one-third to one-half the average taken by all of the households. Virtually no households take one-fourth to one-third the average taken by all of the households, let alone even less.

Some households are in between. They're "medium users" of grid electricity. Medium users take around the average amount of grid electricity.

If the distribution of a utility's households were normal, the medium users would greatly outnumber both the light users and the heavy users. As we've said, the distribution isn't normal. One of the fallouts is that the light users outnumber the medium users. Most households served by a utility take significantly less grid electricity than the overall average. Not as many households as you would expect take around the average amount of grid electricity.[30]

Not that many households are heavy users. But, and this is key, those households that are heavy users tend to take much more from the grid than the overall average. Every utility

30 Janine Morley and Mike Hazas, "The Significance of Difference: Understanding Variation in Household Energy Consumption," ECEEE Proceedings, 2011. "Families living in identically-designed homes use strikingly different amounts of energy... The extent of this variability also implies that there are no "typical" energy-using households within a society... particularly in electricity consumption...and that statistical averages will offer poor representations of actual consumption patterns... The distribution of electricity consumption in the sample of 72 UK homes...shows that only a minority (26%) fall within a roughly comparable 10% above or below the average consumption figure. And, in fact, the majority (60%) fall below average." The author of this book has found that similar distributions characterize American households.

has a surprising number of heavier users and heaviest users at the extreme right of their distribution of households. Just like political extremists, electricity usage extremists exert influence well beyond what their numbers would suggest.

Statistician who drowned

The average is such a common statistic to describe a distribution of whatever. But now you can see how deceptive the average can be. The average, when used to describe electricity trends, masks the rich diversity in households' electricity usage and their utility bills.

It's common to lazily lean on averages. Energy experts routinely cite national annual averages. By doing so, critical regional differences are obscured. Seasonal differences are obscured. Income differences are obscured.

Watch out when people cite averages—about electricity particularly. This undependable statistic can be dangerously misleading.[31]

That seems wrong

A significant majority of households take fewer kilowatt-hours from the grid than the average. For some utilities, 60 to 65 percent of households take fewer kilowatt-hours than the average.

How can this be so? How can it be 60 to 65 percent? That seems wrong. Shouldn't 50 percent of households take fewer kilowatt-hours than the average? The remaining 50 percent then take more than the average. That's better.

Well, no. The simple symmetry of half below the average and half above doesn't fit what actually occurs. The reality is not symmetric. Many more households are below the average than above.

31 Sam L. Savage, *The Flaw of Averages*, John Wiley & Sons, Inc., 2009. "An apocryphal example concerns the statistician who drowned while fording a river that was, on average, only three feet deep."

Whenever someone starts citing average kilowatt-hour usage or average utility bills, say, "Stop. You're exaggerating how much the majority of households use or pay." Don't let them get away with it.

We all thought we could depend on the average to be informative. Sorry to break the news. Statisticians don't call the average the mean for nothing. Averages don't particularly care about us and our ability to understand things.

Graph 8:

Sixty percent of households have electricity costs less than the average; twenty-one percent have less than half the average.

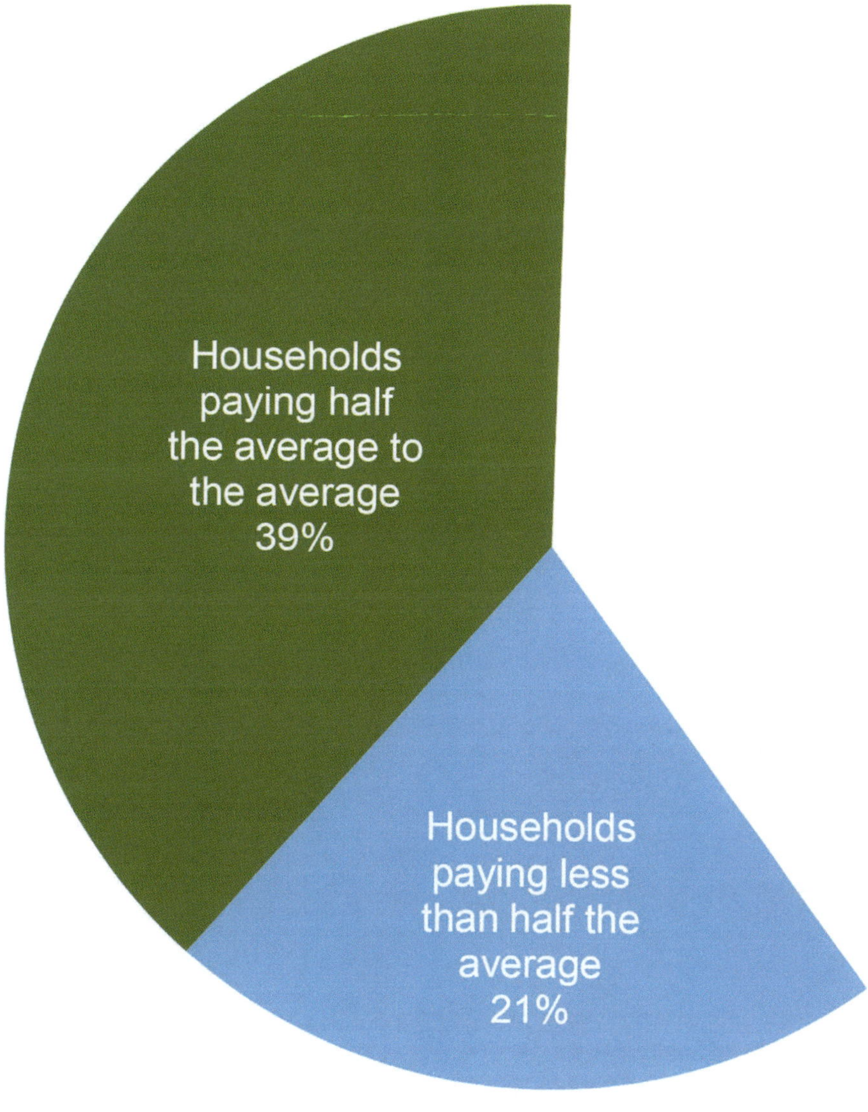

Graph 9:

Many households pay under one dollar daily for electricity, one to two dollars daily, and two to three dollars daily.

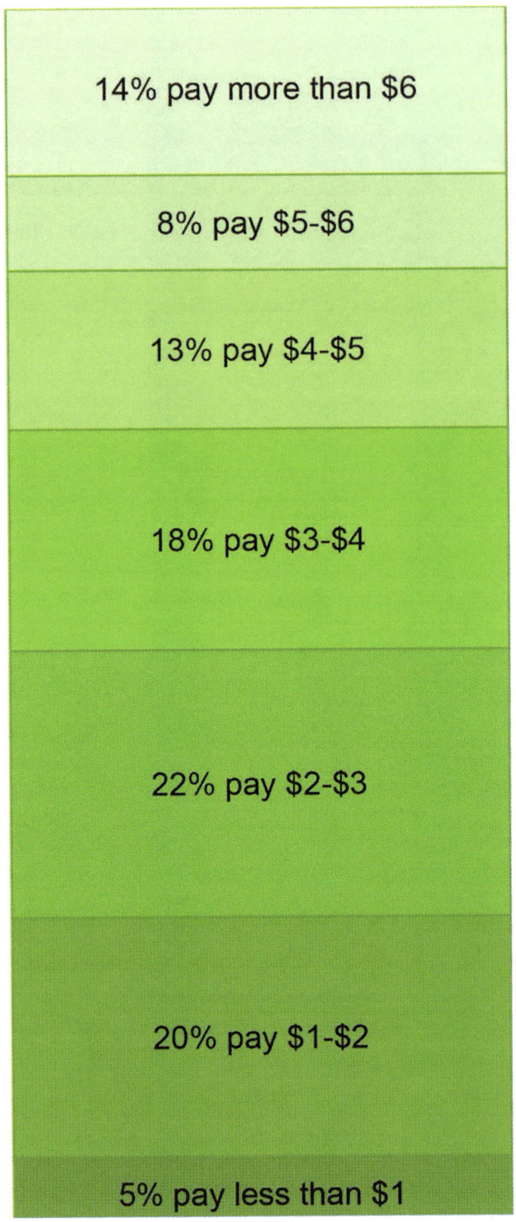

Heavy takers

A minority of all households served by utilities are heavy users of grid electricity. They're the culprits. They're why we don't get the elegant symmetry in the distributions that we expect, neither for electricity usage or utility bills.

Heavy-usage households may be a minority. But they're heavy hitters. They get their way and mess with our statistics.

Even though they're a minority, their max-out usage exerts a firm pull on the overall average. Heavy users pull up the average a lot, well above the usage patterns of the majority.

How strong is this effect? The overall average can be pulled up by so much that 60 to 65 percent of households fall to below it.

Let's illustrate the point by looking at the Ampere family. The amount of electricity the Amperes take is somewhat on the high side, compared with all of the households served by their utility. If the distribution of the utility's households were normal, the Amperes would be above average. But because the distribution is skewed, the Amperes fall to below average.

If the distribution were normal, then 50 percent of households would take less electricity than the average. So the fact that 60 to 65 percent of households take less electricity shows you how skewed up the distribution is.

Broad and broader

Imagine a representative electric utility. We'll call our utility SureVolt Power & Light Company. Their motto is, "You Can Rest Assured with SureVolt."

Households' monthly utility bills average $105. That's a typical average in the United States. On a daily basis, $105 monthly is $3.50.

But SureVolt utility bills vary broadly. That average is not that informative. Of all the households served by SureVolt, very few actually have electricity costs of exactly $3.50 per day. Many have lesser costs than $3.50. Many have greater costs than $3.50. Some households have much greater costs than $3.50, and a few have much lesser costs.

It's like saying the average time that it would take SureVolt customers to run a mile is fifteen minutes. This average is as uninformative as the average is for customers' electricity costs.

The speediest SureVolt customers can run a mile in around five minutes, one-third the average. The slowest SureVolt customers can't run it in twenty minutes; they never finish. Their mile times are infinitely higher than the average. So how useful is that average?

A large number of SureVolt customers have electricity costs around $1.75 daily. These light users are at half the average. Some customers have costs around $10.50. These heavy users are at three times the average or more. So how useful is that average?

Skewed left, skewed right

Graph the statistical distribution of SureVolt customers' mile times. The right-hand tail of slowpokes has way more customers than the left-hand tail of speed demons. That's why the overall average of all customers' times is as high as fifteen minutes. The slowpokes have greater weight on the average than the speed demons because they have greater numbers.

The fifteen-minute average is relatively close to the right-hand tail of slow pokes, twenty minutes or longer. On the other hand, the fifteen-minute average is relatively distant from the left-hand tail of speed demons, five minutes. Because the speed demons' run time is so much less than the overall average, their extreme pulls down the overall average from twenty minutes to fifteen.

We say that the statistical distribution is skewed left toward the left-hand tail. We also say a statistical distribution like this one has a negative skewness.

Now graph the statistical distribution of SureVolt customers' utility bills. The left-hand tail of light electricity users has way more customers than the right-hand tail of heavy users. That's why the average of all customers' electricity costs is as low as $3.50 per day. The light users have greater weight on the average than the heavy users because they have greater numbers.

The $3.50 daily overall average is relatively close to the left-hand tail of light users, two dollars daily or less. On the other hand, the $3.50 daily average is relatively distant from the right-hand tail of heavy users, sixteen dollars daily or more. Because the heavy users' electricity costs are so much greater than the overall average, this extreme pulls up the overall average from two dollars to $3.50.

We say that the statistical distribution is skewed right toward the right-hand tail. We also say a statistical distribution like this one has a positive skewness.

Bridge too far

The statistical distribution of customers' mile times has a barrier created by human behavior that distorts the curve. People seldom run faster than five minutes per mile. People never run much faster than four minutes per mile. So the statistical distribution of run times has nearly zero customers between four and five minutes per mile. It has zero customers less than four minutes. We say that the left-hand tail of the distribution is truncated at four minutes per mile.

Similarly, the statistical distribution of customers' electricity costs has a barrier created by human behavior that distorts the curve. People seldom use so little electricity that their costs are less than one dollar daily.

People never use zero electricity. Even if they did, people cannot drive their bills to zero. SureVolt, like any utility, has a minimum demand or customer charge monthly. Even if a SureVolt customer shuts off the circuit breaker to the house and permanently camps out in the backyard, a SureVolt bill will come in the mail each month. They'll be a charge of around thirty cents daily, including state and municipal taxes and fees.

So a SureVolt customer cannot pay less than thirty cents daily. We say that the left-hand tail of the distribution is truncated at thirty cents daily.

Left versus right

This is a big deal when studying the statistical distribution of customers' utility bills. The difference between the average of all SureVolt customers' electricity costs, $3.50 daily, and the truncation at the left-hand tail, thirty cents daily, is just $3.20. Why is this small difference of monumental importance?

Because this difference is far smaller than that between the average of all SureVolt customers' electricity costs, $3.50 daily, and the right-hand tail of heavy users of electricity, say sixteen dollars daily. The $3.20 difference from the average to the truncation at the left-hand tail is a fraction of the $12.50 difference in the opposite direction, from the average to the right-hand tail. Our statistical distribution is very positively skewed.

Still don't get why this is such a big deal? No one pays less than $3.20 below the average. But plenty of customers pay more than $3.20 above the average. The distribution is especially lopsided.

A number of customers pay more than $6.40 above the average, two times the $3.20 left-hand gap. Some customers pay more than $9.60 above the average, three times the $3.20 left-hand gap.

As a consequence, the average is drawn arithmetically to the right-hand tail, though a relatively small number of customers are at this extreme. Because the heavy users consume so much more electricity than the bulk of SureVolt customers, though heavy users are not that numerous, they pull up the overall average of all customers' electricity costs to $3.50 daily.

We're not normal

This can be illustrated with a contrasting representative utility. We'll call this one GreatWatts Electric & Gas Company. Their motto is, "What Has Great Energy? GreatWatts Does."

GreatWatts has real differences with SureVolt aside from their mottos. GreatWatts has a uniformly normal and perfectly balanced statistical distribution of customers' electricity usage and bills. You'll recall SureVolt has an abnormal and unbalanced statistical distribution of customers' usage and bills.

As we'll see, neither GreatWatts nor any utility can have a normal and balanced statistical distribution of customers' usage and bills. GreatWatts simply can't exist. Its numbers are impossible.

Suppose that the average of the utility bills that GreatWatts sends out in the mails to customers is $105 monthly or $3.50 daily, just like SureVolt. But GreatWatts' customers are normal, at least with respect to the statistical distribution of their usage and bills.

Being normal, there is a perfect balance, a beautiful symmetry around the average. There is one customer whose electricity costs are x cents daily more than the overall average, symmetrically, to balance one customer whose costs are x cents less than the overall average.

With an average of $3.50 daily, there is one GreatWatts customer paying three dollars for every one GreatWatts customer paying four dollars. One customer pays $2.50 for every one customer paying $4.50. One customer pays two dollars for every one customer paying five dollars. One customer pays $1.50 for every one customer paying $5.50. It's all perfectly balanced.

Follow me here. There is a rewarding end to this, I promise you. One GreatWatts customer pays one dollar for every one GreatWatts customer paying six dollars. One customer pays fifty cents for every one customer paying $6.50. One customer pays nothing for every one customer paying seven dollars. Oops. That doesn't sound right. How can a household pay nothing for its electricity?

Want to hear something that really sounds wrong? For GreatWatts to have a normal and perfectly balanced statistical distribution of customers' usage and electricity costs, it would need to have one customer paying minus fifty cents for every one customer paying $7.50. Something is evidently very wrong. No household pays minus fifty cents daily for grid electricity.

Every utility in the United States has residential customers paying electric bills that average out to be $6.50 per day, seven dollars per day, even more. No utility has residential customers paying nothing per day, minus fifty cents per day, even less. The prosecution rests.

We have proven something really important in electricitynomics. No utility's distribution of households is normal, perfectly balancing the highs and lows. Everyone is, well, abnormal and unbalanced. The distributions are all positively skewed.

That's heavy, man

The more numerous the heavy users are among a utility's residential customers, the more the distribution will be distorted from a symmetric normal distribution. The more extreme the usage of electricity is by the heavy users, the more the distribution will be distorted.

The distributions of many utilities look nothing like the classic bell curve. Heavy-usage households are numerous in these utilities' areas. Their heavy users take extreme amounts of electricity. The curve, rather than bell-shaped, looks like the dog sat on one side of it.

One in ten of a utility's residential customers can be heavy users, 10 percent. It doesn't seem that numerous. But they sure can do major damage to a utility's distributions.

Suppose each heavy-usage household takes at least 2,250 kilowatt-hours monthly. We're smoothing out all of their usage over the year. Their spring and fall usage falls well short of 2,250 kilowatt-hours. Their summertime usage easily makes up the shortfall.

That's impressive, 2,250 kilowatt-hours monthly. It's equivalent to seventy-five kilowatt-hours daily. In hourly terms, it's around three kilowatt-hours.

No question, that's big-time usage of the grid for a single residence. Heavy-usage households each take about double what the average American household takes.

Heavier still

Suppose one in three of these are heavier-usage households. We're not talking about a lot of homes here. One in three, the heavier user subset, of one in ten, the heavy user set, equals one in thirty. So, one in thirty of all a utility's residential customers are heavier users. In percentage terms, that's only about 3 percent.

Suppose each heavier-usage household takes at least three thousand kilowatt-hours monthly. We're again smoothing out all of their usage over the year.

That's really impressive, three thousand kilowatt-hours monthly. It's equivalent to one hundred kilowatt-hours daily. In hourly terms, it's around four kilowatt-hours.

No question, that's very big-time usage of the grid for a single residence. Heavier-usage households each take nearly triple what the average American household takes.

She's so heavy

Then there's the extreme of the extreme. They're on the outer fringe in household use of the grid. They're the heaviest-usage households.

Suppose one in ten of the heavy-usage households, one in three of the heavier-usage households, are the heaviest users. We're talking about very few homes now. One in ten, the heaviest user sub-subset of one in ten, the heavy user set, equals one in a hundred. So, one in a hundred of all a utility's residential customers are the heaviest users. In percentage terms, that's only 1 percent.

Suppose each heavier-usage household takes at least 3,750 kilowatt-hours monthly. We're again smoothing out all of their usage over the year.

That's crazily impressive, 3,750 kilowatt-hours monthly. It's equivalent to 125 kilowatt-hours daily. In hourly terms, it's around five kilowatt-hours.

No question, that's super big-time usage of the grid for a single residence. Heaviest-usage households each take more than triple what the average American household takes.

Now let's reassemble the heavy set of households, that 10 percent of a utility's residential customers that takes more than 2,250 kilowatt-hours from the grid smoothed over the year. The heavy set is made up of three subsets. There's the subset of heavies that don't take as much from the grid as heavier users or heaviest users. There's the subset of heavier users. And there's the subset of heaviest users.

Reassembled, the heavy set of households averages 2,850 kilowatt-hours monthly. On a daily basis, that's ninety-five kilowatt-hours. These households are power-hungry, no question. But because seven in ten of them are neither in the high-end heavier subset or real-high-end heaviest subset, the average is weighed down by the less extreme heavies. Even within the heavy set, we see a skewed distribution.

Heavies are big

Although the heavy-usage households are not numerous, when compared with all of the residential customers of the utility, they exert a disproportionate effect on the summary statistics for all customers. To show the outsized effect of the heavies, let's take our example further.

Suppose that for all of a utility's residential customers, whether heavy-usage or not, the overall average take from the grid is one thousand kilowatt-hours monthly, thirty-three kilowatt-hours daily. Then the math shows that the heavies take 28.5 percent of the total take by all residential customers.

This is so, though the heavy users are so few. They are just 10 percent of all residential customers. All of the other customers, the 90 percent that are non-heavy, take just 71.5 percent of the total take.

What if the utility in question has what's called a flat-rate structure? All customers pay the identical amount per kilowatt-hour, whether they are heavy or non-heavy. If so, the heavies pay 28.5 percent of residential utility bills in dollars. The non-heavies pay 71.5 percent, 2.5 times as much as the heavies, though the non-heavies are nine times as numerous.

The same goes when utility regulators set utility rate increases. The heavies pay 28.5 percent of the rate increase, and the non-heavies pay 71.5 percent, spreading the increase thinly among nine times as many households.

More skewed than others

Utilities' distributions of their residential customers vary. Some utilities have distributions that are more positively skewed than those of other utilities. Some utilities have distributions with unusually high "kurtosis" or peakedness.

There are three key contributing factors that determine whether a utility's distributions are very skewed and peaked. First, do the utility's customers have widely varying incomes? Second, does the utility's area have high population growth? Third, have the utility's regulators set the utility rate structure as an inclining block, a declining block, or a flat rate?

A utility has a flat-rate structure if households that take light quantities of electricity, medium quantities, and heavy quantities all pay about the same per kilowatt-hour. There is no quantity discount, nor is there a quantity penalty. The household that takes five thousand kilowatt-hours in a month pays around ten times more than the household that takes five hundred kilowatt-hours.

A second utility has an inclining block-rate structure. For the households it serves that take heavy amounts of electricity, they pay significantly more per kilowatt-hour than households that take light amounts of electricity, with medium-usage households in between.

The household that takes five thousand kilowatt-hours in a month pays much more than ten times what the household that takes five hundred kilowatt-hours pays.

A third utility has a declining block-rate structure. For the households it serves that take heavy amounts of electricity, they pay significantly less per kilowatt-hour than households that take light amounts of electricity, with medium-usage households in between. The household that takes five thousand kilowatt-hours in a month pays much less than ten times what the household that takes five hundred kilowatt-hours pays.

If a utility's customers have widely varying incomes, as in Washington, DC, for example, the utility's distribution will be very positively skewed and have high kurtosis, all other things being equal ("ceteris paribus," for Latin fans). If, on the other hand, there is relatively equal income distribution, as in much of the Midwest, for example, the bill distribution will not be as distorted from the classic bell curve.

If a utility has rapidly growing population, as in much of the sunbelt before the 2008–2012 recession, the utility's distribution will be very positively skewed and have high kurtosis, all other things being equal. If, on the other hand, population is relatively flat, as in much of the northern tier of the country, for example, the distribution will not be as distorted from the classic bell curve.

If a utility has an inclining block-rate structure, as in California, for example, the utility's distribution will also be very positively skewed and have high kurtosis, all other things being equal. If, on the other hand, the utility has a declining block-rate structure, as in much of the Midwest, for example, the bill distribution will not be as distorted from the classic bell curve.

The supremes

Heavy-usage households, especially the heavier users and the heaviest users, take a lot from the grid. They pay for it.

These households in aggregate can be just 10 percent of all of a utility's residential customers, a small minority. But they can take over 20 percent of the electricity and pay over 30 percent of the dollars.

The heavier users can be just 3 to 5 percent of all customers, a tiny minority. But they can take over 12.5 percent of the electricity and pay over 20 percent of the dollars.

The heaviest users can be just 1 to 2 percent of all customers, a real tiny minority. But they can take over 7.5 percent of the electricity and pay over 10 percent of the dollars.

This becomes more tangible in an example. A utility, Resistance Electric Company, has a million residential customers. To keep its end of the grid in good working condition, Resistance Electric needs to collect a billion dollars annually from its residential customers. Its heaviest users number just twenty thousand, 2 percent of the million. Even though they're a real tiny minority, they pick up 10 percent of the tab, or $100 million, all by themselves.

Per customer, the heaviest users pay five thousand dollars annually. That comes out to be $417 monthly or $13.70 daily.

The remaining 980,000 residential customers of Resistance Electric pick up the remaining 90 percent of the tab, or $900 million. Per customer, the customers who are not heaviest users pay $918 annually. That comes out to be $76.50 monthly or $2.52 daily.

Resistance Electric is able to raise the cash to run its end of the grid. It has twenty thousand heaviest users each paying five thousand dollars. It has 980,000 other customers each paying $918. That gives you an overall average of one thousand dollars per customer. The heaviest users pay 400 percent above the average. The other customers pay 8 percent below the average.

Let's look at the entire heavy set of residential customers. Add up the twenty thousand heaviest users, the thirty thousand heavier users, and the fifty thousand other heavies. The heavy user set in aggregate comes to 100,000 of the million customers of Resistance Electric, 10 percent.

The heavy user set pays $300 million of the billion dollars needed to keep up the grid. The non-heavy set pays the remaining $700 million needed. How does this calculate out? The customers in the heavy set pay an average of three thousand dollars each. The customers in the non-heavy set pay an average of $777 each.

Still, the overall average is one thousand dollars per customer. Heavy-usage households pay 200 percent above this average. Non-heavy households pay 22 percent below this average.

Nearly free power for 40 percenters

Households that aren't heavies in their electricity take pay $777 each per year for what they take. There are many households in this category, 70 percent of all of the utility's residential customers. That $777 is the simple simplistic average.

Some of the 700,000 non-oversize households pay more than $777. An offsetting number pays less than $777. Over one-third of all residential customers pay less than $777. On a monthly basis, $777 annually is just sixty-five dollars. On a daily basis, $777 annually is just $2.13.

However, it's crazier than that. Considerably more than 350,000 customers pay less than the simple simplistic average. Once again, it's because of electricity skewed. There are again many more utility customers below the average than above.

The distribution must be skewed. It has no other choice. Some non-heavy households pay $777 more than the $777 average. Some pay even more than that. But no non-heavy households pay $777 less than the $777 average. They would be getting their grid electricity for free. Nobody pays even less than that. They would be getting their electricity for better than free. The utility would be paying them to take electricity from the grid.

Perhaps 40 percent of the million utility residential customers, 400,000 households, pay less than $777 annually. Stop and consider what this means. Hundreds of thousands must pay less than six hundred dollars annually. On a monthly basis, that's fifty dollars. On a daily basis, that's $1.67. For them, grid electricity is not free but not far from free.

Reddy's Recharges

1. Every utility's distribution of households' electricity usage and bills is skewed rather than normal.

2. This skewedness comes from the extreme usage of electricity by the minority of all households that are heavy users.

3. Some utilities have distributions that are more skewed than those of others because of their area's population growth, income disparity, and the rate structure that regulators have set.

"I'm a social scientist, Michael. That means I can't explain electricity or anything like that, but if you ever want to know about people I'm your man."

How We *Use* Grid Electricity

On why grid electricity's usage, costs, and value to the public are best represented as skewed statistical distributions and how usage is best forecasted

PART III

Part III. How We *Use* Grid Electricity

Chapter 6: Our Use Is Not Normal But Skewed

Chapter 7: What's Happening to Our Use?

Proof, so what?

I showed it in so many different ways. I looked at whether electricity costs are high or moderate and whether they consistently increase or are stable. The answer came back the same each and every time. All of the analyses were rigorous. All of the results were clear-cut, indisputable. What more proof do you want?

There's just no legitimate way to show grid electricity costs have been high and ever-increasing. Try to. It can't be done. Electricity costs are definitively moderate and stable.

So I figured that's it, game over. People would say, "Okay, we get it. Electricity costs are actually moderate and stable. We didn't know Steve. Now we do. From now on, when utilities propose investments to modernize or expand the grid, we promise we'll evaluate it on the merits. There'll be no more no's, just as a reflex. No more rejecting practically anything utilities propose. No more relying on mistaken beliefs about electricity costs."

Of course, that's not what happened. The reaction to the June 2011 report, replete with proofs, was ho-hum. Utility leaders responded similarly. No news here. Dog bites man. We knew.

Utility leaders know electricity costs are moderate and stable. But they sure feel uncomfortable saying it. This odd discomfort springs from peculiar concerns. Even when armed with slam-dunk proofs, utility leaders are uncertain how the facts about electricity's cost can be brought to bear to help.

Utility regulators, on the other hand, don't know what to think. Most suspect that electricity costs aren't that high and aren't always increasing. After all, regulators set electricity costs with their own hands, their votes in rate cases. They should understand trends in electricity costs as well as anyone.

However, a weird dynamic is at work. Groups opposing utilities, with various agendas, constantly allege electricity costs are anything but moderate and anything but stable. And, as we said, utilities generally respond to the criticism with little more than stoicism. Rarely do utilities try to dispel the notion that opponents push, that electricity is expensive and becoming more so all the time.

Actually, the situation is worse. Utilities often make apologies when proposing investments to modernize or expand the grid's plant and equipment. In part to show empathy with utility customers, they go to great lengths to express their regrets that electricity costs will go up if regulators accept the proposals.

That the cost increase is likely to be marginal for customers is noted but with little conviction. That the cost increase is likely to be a fraction of the increased value for customers is noted but with considerable caution.

Utilities are wary of appearing insensitive to any increase in customers' costs. They're also wary of appearing aggressive about "selling" the great value of their service. You can understand why. It doesn't take much for opponents to hammer utilities for being insensitive about cost increases or being aggressive about selling. Being insensitive, being aggressive, these are no-no's.

Hearing the allegations and hearing little or no rebuttal, the media and public naturally take the allegations as the facts. Electricity costs must indeed be high and consistently increasing. Why else would utilities be almost silent on the matter? You can't blame the media and public for feeling this way.

Utility regulators may understand deep down that their control of electricity costs has been effective. But the media and public are convinced otherwise.

It is the common belief, however mistaken, that drives the regulatory negativity to utility proposals. It is this belief that gives us the dead-on-arrival attitude in utility regulation. "Nice proposal, utility. But customers just can't afford it given how increasingly expensive electricity is."

If I was going to turn around this sorry condition, in which utility regulation behaves as if electricity costs are high and ever-increasing when they're not, I would need to do something very different. I would need to do something much more effective than put out a report with proofs.

Drilling for energy data

My journey thus took me in search of more data. I looked for something else, something more compelling, and something that would really make the case.

We went back to the Energy Department data. But this time, we drilled deeper, much deeper.

That's when we found the Energy Department's Residential Energy Consumption Survey. It has a strange acronym, RECS, as in "wrecks." I did find that exploiting RECS' wealth of information helped us wreck the common misimpressions about electricity costs.

Like other energy experts, I had always known RECS was there. But I had paid it little heed. RECS always seemed focused on tracking and trumpeting the nation's progress in energy efficiency ever since the presidency of Jimmy Carter.

This neglected database shows what American households pay for electricity in extraordinary detail. The detail enables researchers to characterize the diversity and distribution of electricity usage and bills among the 120 million households nationally.

RECS' accuracy is unequaled anywhere else. The Energy Department takes an awfully long time getting the results of RECS out. This is largely because the department takes extreme measures to double-check the utility bills of survey participants. For 2009, the latest year that is reported, over twelve thousand households participated. The department double-checked the utility bills of all twelve thousand households, verifying the bills with participants' utilities.

Then drilling for consumer data

It was July 2011. We were digging through the RECS, looking for riches. It was then when my team really went into overdrive. We happened onto an even more extensive survey of what American households pay for electricity. The survey had been conducted continuously for decades by the Labor Department and was totally up to date. We couldn't believe our good fortune.

It's crazy. I had never used the Consumer Expenditure Survey. I had never even heard of any energy expert using the CES.

Each year since the presidency of Ronald Reagan, the Labor Department has been surveying some thirty-five thousand households. The end product is incredible. The CES tells us how much American households spend on every category of goods and services, including grid electricity. CES tables report how household expenditures vary by income, region, and other characteristics. CES microdata allow research teams like my own to delve much deeper.

Both sets of microdata, of the RECS and CES, are extraordinarily rich seams of intelligence. The detail on how American households differ in their usage of electricity, and how their bills differ, literally made us giggle with excitement.

Our statistical programs were having a field day. It was then, during the summer of 2011, that the team made its most monumental discovery. For the first time, our confidence grew that our central hypothesis was spot on.

Plot the distribution of households' electricity usage. Or plot the distribution of their bills. Do it for all US households or any demographic slice. The shape of the distribution always looks the same.

The shape is not normal. It is not a normal distribution. The shape looks nothing like the classic bell curve that all high school students learn. The shape is a skewed distribution. With a long right-hand tail, it's off-centered.

So, our hypothesis is true. Electricity is skewed!

A gusher, utility customer bill data

Could we really nail it? Could we prove for certain that electricity is indeed skewed?

Then it came. The breakthrough came. It was October 2011. First one utility allowed us in. My team was given permission to analyze the utility's customer bill database as much as we wanted. We sure did want to.

Then two other utilities allowed us in. Just three months later, it was now January 2012. My team was given unprecedented access to customer bill databases of three utilities.

The three utilities together are the grid electricity providers to around 4 percent of the United States, over 5 million households and businesses. Wow! This was a gigantic sample. We would now be able to analyze nearly 350 million monthly customer bills. That's how many bills were in the three databases, bills rendered to households and businesses in the South and Midwest over the past few years.

We had no idea as to the identity or address of any of the households in the databases. This was a critical condition of our access, appropriately required by the three utilities. Utility customers' privacy could not be compromised. There was no possibility of a breach of confidentiality even inadvertently.

What we did have access to was some mighty key statistics. The team could see, for each customer, for each month, their monthly electricity usage in kilowatt-hours and monthly electricity bill in dollars and cents.

The RECS and CES microdata were great, to be sure. But now we had real data from the utilities themselves. We had lots and lots of real data. It was a gusher. We were open for business to validate our central hypothesis and other hypotheses. We would now be able to characterize Americans' usage of electricity and their bills as never before done, as never before attempted.

Blowing past averages

What was so awesome about gaining access to the three utility customer bill databases? Isn't data on Americans' electricity usage and bills already out there? Hasn't this data already been analyzed to death?

Sure, the Energy Department reports on Americans' electricity usage and bills for all to see. The reports are pretty good. The reports are even detailed enough so you can find out customers' average usage and average bill, for each utility, for each month. They're broken out so you can see the averages for households ("residential customers") and businesses (separately for "commercial customers" and "industrial customers"). Isn't that enough data for any practical purpose?

No, it's not. The Energy Department data are so average. Actually, that's the point. The Energy Department data are averages. Excepting the RECS, that the data are averages is a critical limitation.

When all we have are customers' average electricity usage and average electricity bills, we are blind to how customers' usage and bills vary. When all we have are average usage and bills, we can't see the broad and surprising diversity in Americans' relationship with our electrical grid.

What makes the Energy Department's RECS and Labor Department's CES so insightful and so unique is their microdata detail on individual households' electricity usage and bills. But as great as RECS and CES are, they have their own limitations. RECS is reported only sporadically. Its most recent year is 2009. Before that, the most recent year is 2005. CES is more up to date. But CES has only electricity bill data. It leaves out electricity usage data.

As neglected as RECS and CES have been by energy experts, to understand Americans' relationship with the grid, utility customer bill databases have been left virtually untouched. Don't get me wrong. The databases are heavily used by the utilities themselves for financial and accounting purposes. But not to assess customer trends in electricity usage and electricity bills. What my research team did was unique. Even more unique, we looked for customer trends in the utility customer bill databases across multiple utilities in multiple regions.

It isn't surprising that utilities have not been using their own databases in recent years to evaluate customer trends in electricity usage and electricity bills. It has been that politically incorrect in recent years for utilities to do anything that that might look like marketing.

By January 2012, my team could run statistics on nearly 350 million monthly bills. We felt like explorers landing on the shores of some faraway place not seen before by man. And what we discovered, what we saw previously unseen by anyone in the energy field, felt just as spectacular.

Now for Part III

In Chapter 6, "Our Use Is Not Normal But Skewed," you'll read about how we started seeing the same kind of statistical distributions, skewed ones, everywhere we looked. The public's electricity usage, it's skew-distributed. The public's utility bills, they're skew-distributed. Utility blackout durations, they're skew-distributed. The public's blackout costs, they're skew-distributed as well.

In Chapter 7, "What's Happening to Our Use?" you'll see how this research sparked new insights about how electricity usage grows in twenty-first-century America. The team could now answer questions about electricity sales growth, or the lack of sales growth, that were heretofore confounding mysteries to even seasoned energy experts.

The journey began innocently enough with an itching curiosity about electric rates and bills. Part I told the story of these first tentative steps. Soon after, the journey was transformed into a wondrous tour, full of surprises. The team had its first inklings about electricity's value. Part II told the story of these surer steps. Part III will recount the growth spurt, when we experienced a surge of discoveries once we were able to lay our hands and computers on hundreds of millions of utility records.

Electricity sales growth is almost entirely from household formation. Existing homes that are decreasing electricity use are netting those that are increasing use. New homes, in contrast, add large increments of use. New homes also tend to have large interior volume to cool and heat. So, despite energy efficiency features, they take more electricity from the grid than existing homes.

Chapter 6: Our Use Is Not Normal But Skewed

Majority of households in distribution's hump; heavy taker minority in right tail; due discrimination

Skew rules

Skewed distributions rule! More specifically, positively skewed statistical distributions rule!

Who are these kings of so many realms of electricitynomics? Skewed distributions are powerful. They're beautiful. They have graceful asymmetric curves. A majestic hump sits on a throne to our left. An asymptotic tail, long and slender, extends out far to the right.

Imagine if a skateboard park were shaped like a skewed distribution. The curves would intimidate all but the most daring. The slope from the hump down to the tail is far too steep for most boards. It is especially steep at the terrifying point of inflection midway down.

Some test the slope anyway. The daredevils are last seen careening at max speed along the tail to our right and then into oblivion.

The skewed distribution may be a promising template for skateboard park design. For our purposes, skewed distributions vividly illustrate the dynamics of electricity's supply and usage and where they're heading.

Majestic hump

The majestic hump on our left is one of the two characteristic features of a positively skewed statistical distribution. The hump signifies that many constituents of the larger group have something very much in common. They're our first key subset of the larger group.

The hump could signify the subset of a utility's residential customers that are light users of grid electricity. Many households fit this category. It could signify the subset that pays low utility bills. Again, many households fit this category. It could signify the subset that pays low bill increases whenever utility regulators raise utility rates. Again, many households fit this category.

The hump could signify the subset of a utility's power outages that are brief disruptions of customers' electricity service. Many outages among all outages fit this category. It could signify the subset of briefly disrupted utility customers whenever there is an extensive power outage. Many households in extensive outages fit this category.

What do the constituents in the hump subset have in common? Whatever is being measured, what they have in common is that their number along this metric is relatively low.

If the metric is how much grid electricity households take, the households that take little are in the hump. If the metric is what households pay in utility bills for their take, the households that pay little are in the hump. If the metric is what households pay for rate increases, the households that pay little for increases are again in the hump.

If the metric is the average duration of customers' service disruptions among all of a utility's outages, the outages that have a short average duration are in the hump. If the metric is the duration of customers' service disruptions for a single extensive outage, the households that luckily experience a comparatively short disruption are in the hump.

Graph 10:

Statistical distribution of households' electricity costs is skewed.

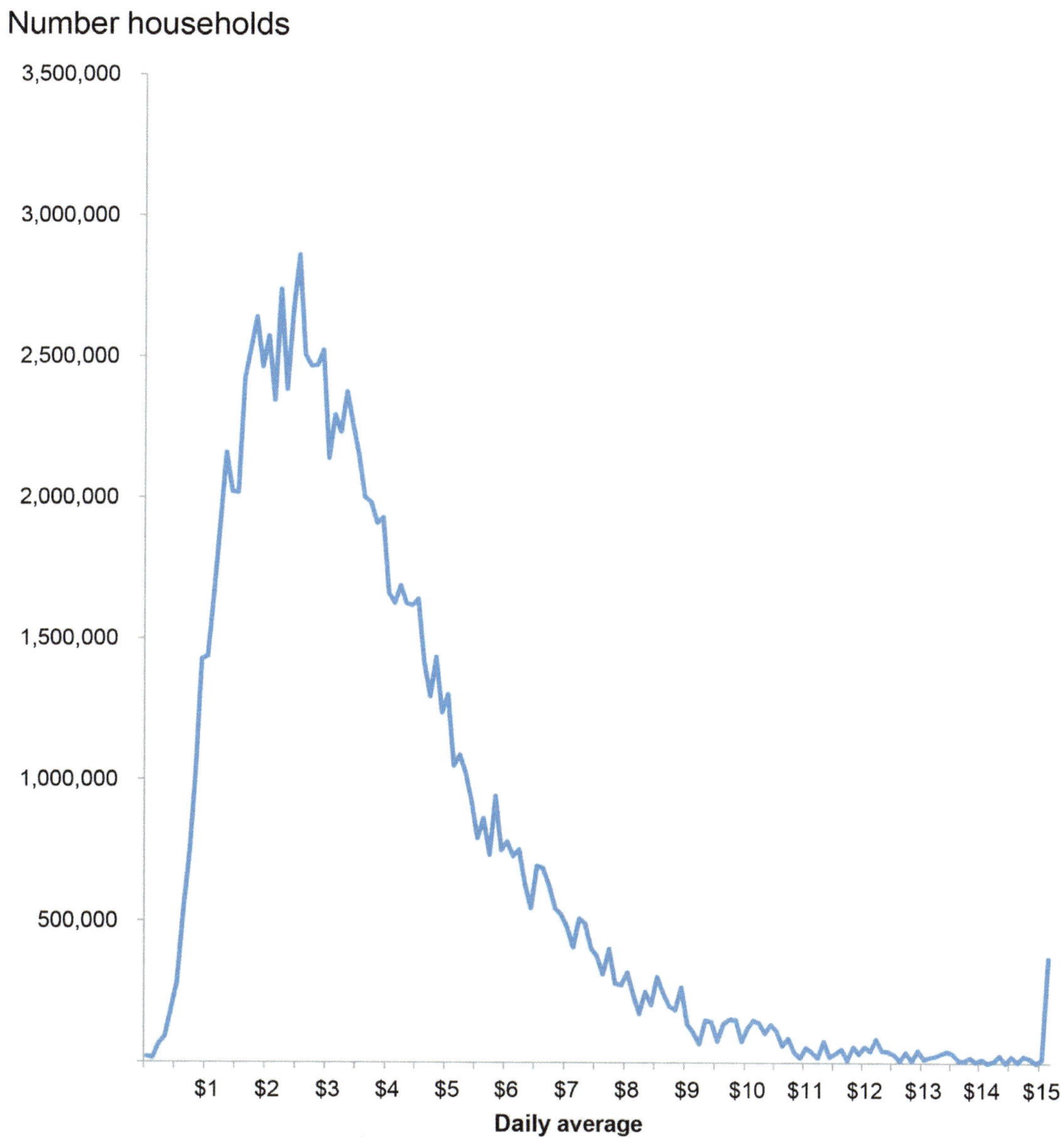

Graph 11:

Bottom quintile of households' electricity costs is under two dollars while top quintile is over five dollars.

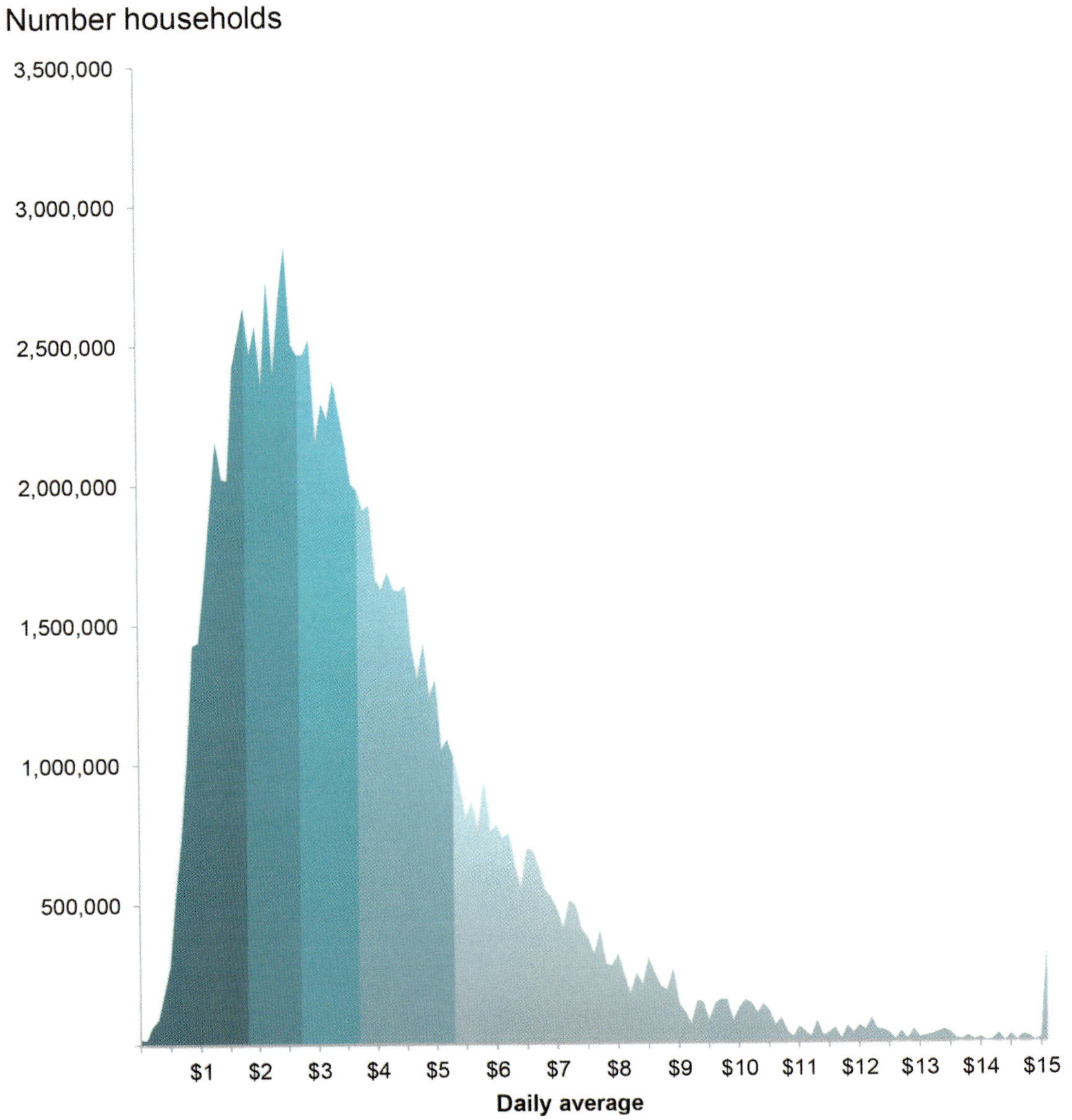

Graph 12:

Skewed statistical distributions of electricity costs are highly seasonal.

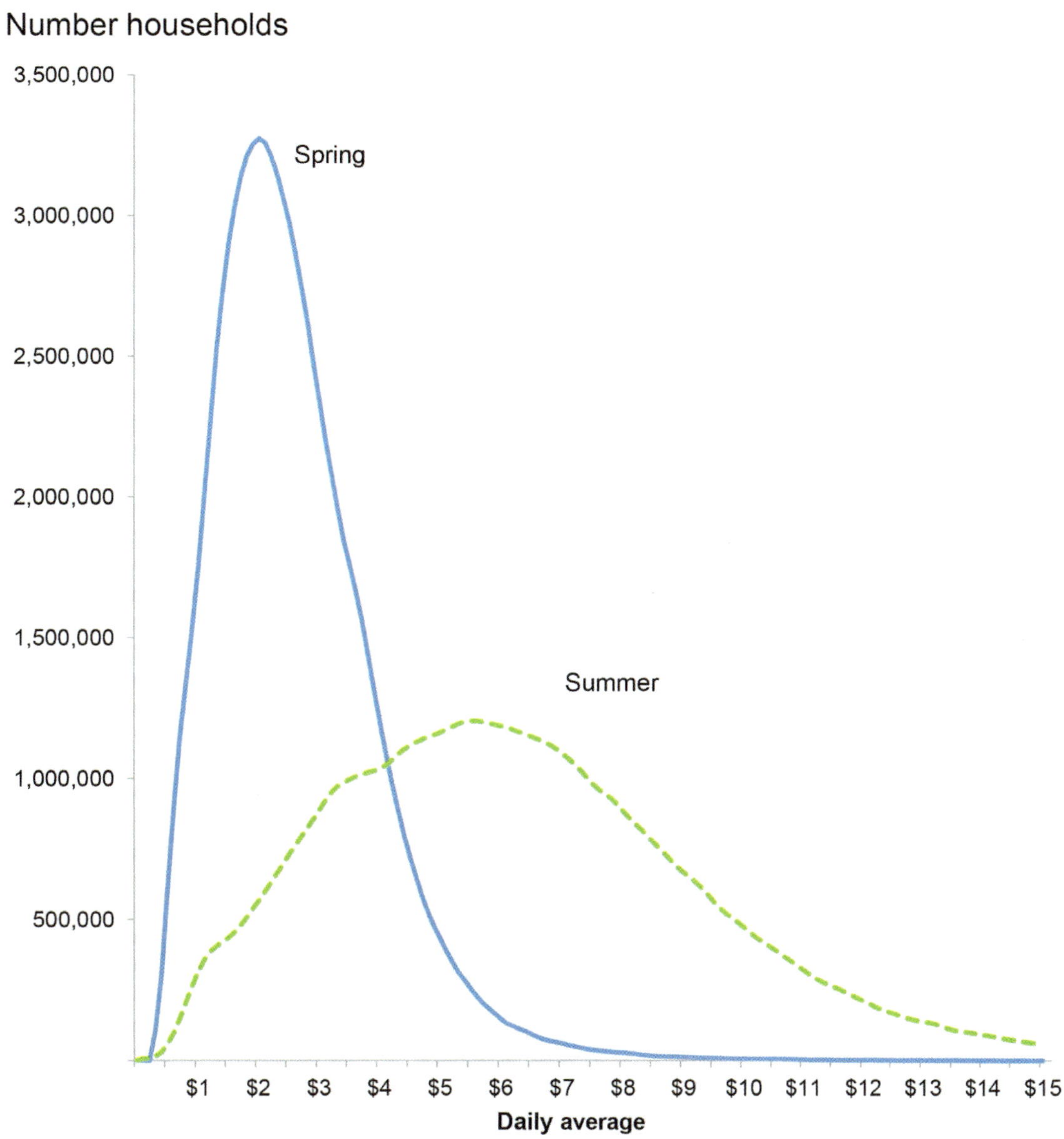

Clumped in the hump

The hump has the majority of the constituents of a group. When discussing a distribution of a utility's customers, the hump has the majority of the customers. When discussing a distribution of a utility's power outages, the hump has the majority of the outages.

The majority of constituents within the hump are clumped together. Whatever is being measured, all of the hump's constituents vary little in their number along this metric.

If the metric is how much grid electricity households take, the households within the hump vary little in how many kilowatt-hours they take. If the metric is what households pay in utility bills for their take, the households within the hump vary little in how many dollars and cents per day they pay. If the metric is what households pay for rate increases, the households within the hump vary little in how many cents per day they pay for increases.

If the metric is the average duration of customers' service disruptions among all of a utility's outages, the outages within the hump vary little in hours and minutes of average duration. If the metric is the duration of customers' service disruptions for a single extensive outage, the households within the hump vary little in the hours and minutes they are disrupted.

So, within the hump, there's a majority of all of a group. The hump subset includes the group's constituents with a low number of whatever is being measured. These constituents vary in their low number of whatever is being measured, but the variance is rather limited.

Hump's peak

The hump has a peak. While the hump is a majority subset of a group's constituents, the "mode" is a big sub-subset of the group's constituents within the hump. As in a scoop of vanilla ice cream sitting atop a slice of hot apple pie a la mode, the mode sits atop the hump.

As clumped as the hump constituents are, many of these constituents have the exact same number of whatever is being measured. They could be the households that all take the exact same amount of kilowatt-hours. The mode could be twenty-five kilowatt-hours daily. They could be the households that all pay in utility bills the exact same amount for what they take. The mode could be $2.50 daily. They could be the households that all pay the exact same amount for rate increases when utility regulators set higher rates. The mode could be fifteen cents daily.

The mode could be the power outages that all have the same average duration. The mode could be forty-five minutes. It could be the utility customers that all have the exact same length of disruption in an extensive outage. The mode could be two hours.

The mode has no competition. More of the group's constituents have the mode than any other number of whatever is being measured. If the metric is households' kilowatt-hour usage, no other usage amount is as common as the mode amount. If the metric is households' utility bills, no other bill amount is as common as the mode amount. If the metric is households' extra payment from a utility rate increase, no other extra payment is as common as the mode amount.

If the metric is a utility's power outage average duration, no other average duration is as common as the mode amount. If the metric is households' length of disruption in an extensive outage, no other length of duration is as common.

The median is the message

Any group has an overall average. Whatever is being measured about the group, you can do the math and come up with an average. The same applies for positively skewed statistical distributions. Yet the average for skewed distributions is, well, skewed. The average or "mean" to geeks is higher than you expect. Skewed distribution means aren't nice.

Just like the mode, the mean is a measure of "central tendency" for a statistical distribution. But for a skewed distribution the mean isn't all that central. Nor is the mode, for that matter.

What's messing with these usually reassuring stats, the mean and the mode, of skewed distributions? Why don't they explain the dynamics of groups when the statistical distribution is skewed? The culprit is our second key subset of the larger group, as we'll discuss next. The culprit is the asymptotic tail.

The "median" is a better medium, or measure of central tendency, when the distribution is skewed. Find the spot in the group where precisely half of the group has a higher number of something and the other half has a lower number of the same something. That's it, the median. When the distribution is skewed, the median is the message.

We've touched on three measures of central tendency for statistical distributions whether they are skewed or not. When the distribution is skewed, the mode is surprisingly low, at the hump's peak. The mean (the average) is surprisingly high, well to the right of the hump. However, the median is located in a good place. Situated between the mode on its left and the mean on its right, the median is fairly descriptive of the dynamics of the group.

Asymptotic tail

The asymptotic tail on our right is the other characteristic feature of a positively skewed statistical distribution. With a tail on the right and a hump on the left, skewed distributions are easily distinguishable.

The constituents within the tail, a minority of the larger group, have something in common, just like the constituents within the hump, the majority, have something in common. We saw that the hump majority has a low number of whatever is being measured and the constituents within the hump are clumped closely together. In contrast, the tail minority has a high number of whatever is being measured and the constituents within the tail are thinly spread out.

The asymptotic tail could signify the subset of a utility's residential customers that are heavy users of grid electricity. It could signify the subset that pays high utility bills. It could signify the subset that pays high bill increases whenever utility regulators raise utility rates.

The tail could signify the subset of a utility's power outages that are lengthy disruptions of customers' electricity service. It could signify the subset of lengthily disrupted utility customers whenever there is an extensive power outage.

The few constituents within the tail have something in common as do the more numerous constituents within the hump. Whatever is being measured, what tail constituents have in common is that their number along this metric is relatively high.

If the metric is how much grid electricity households take, the households that take a lot are in the asymptotic tail. If the metric is what households pay for their take, the households that pay a lot are in the tail. If the metric is what households pay for rate increases, the households that pay a lot for increases are again in the tail.

If the metric is the average duration of customers' service disruptions among all of a utility's outages, the outages that have a long average duration are in the asymptotic tail. If the metric is the duration of customers' service disruptions for a single extensive outage, the households that unluckily experience a comparatively long disruption are in the tail.

Tail wags the dog

The constituents in the tail are a minority but highly influential nonetheless. If not for them, the mean (the average) would be in the center of the action, not off to the side. If not for them, the mean, the mode, and the median would be perfectly aligned and representative of what the whole distribution is really about. The distribution would be normal, not skewed. It would be easier to interpret what's been happening to the larger group and what's going to happen.

A skewed distribution isn't as transparent as a normal distribution. When a distribution is normal, you can quote an average and be done with it. The average will tell you pretty much all you need to know. No such luck with a distribution that is skewed.

That long and slender tail means that its constituents are far out there. Tail members are extremists. Their number, in terms of whatever is being measured, is so much higher than the bulk of the larger group. Counting them in pulls the mean from the mode and the median. Count them in and the mean gets pulled a distance from the mode and the median. The gap can be so great that the mean loses its meaningfulness, even though those in the tail are so few.

We offer up another example to show what the mean means and doesn't. A utility has one hundred power outages in a year. This seems like a lot. But wait. Fifty of the outages have an average duration of ten minutes. Ten minutes isn't so bad. Forty of the outages are thirty minutes. Thirty minutes isn't great but bearable. The remaining ten outages are twelve hours or seven hundred and twenty minutes. Okay, an outage that long can be tough.

Know what the mean of these hundred power outages is? Do the math. The mean comes out to eighty-nine minutes. Let's call that an hour and a half. Consider the implications. This mean duration is at least three times longer than 90 percent of the utility's outages.

Let's illustrate what the tail, with its 10 percent of the utility's outages, did to mess with the mean. Calculate the mean of just those 90 percent of the outages, the ones not in the tail. The non-tail mean is only around nineteen minutes. That's way lower than the eighty-nine-minute mean when you count all hundred outages, the ninety not in the tail and the ten troublemakers in the tail.

Here's another way to see how meaningless the mean is for the skewed distribution in our example. The mean duration is eighty-nine minutes. None of the hundred outages lasted anywhere close to eighty-nine minutes. Half the outages lasted around one-ninth as long. Ninety percent of the outages lasted around one-third as long or less. The remaining 10 percent of outages lasted eight times longer. The mean is a good-for-nothing.

Tastes and behaviors

Americans' tastes and behaviors are diverse. We take pride in treading individual paths. Though varied, our paths have discernable patterns. Millions hunt while millions more would never pick up a gun. Millions pig out while millions more would never stray from a strict diet.

Americans' tastes in electricity and behaviors using it are also diverse. We don't notice the individual paths we tread. How different they are. Yet here, too, our paths have discernable patterns.

Americans' tastes and behaviors are affected by tipping points, herd mentality, transaction costs, institutional factors, panics, fears, greed, complacencies, fashion of the moment, peer pressures, conspicuous consumption, obstacles to switching, envy, ego, selflessness, ambition, laziness, care, carelessness, love, hate. This is true whether we are talking about our tastes and behaviors generally or our tastes in electricity and behaviors using it.

We live our lives almost entirely within interior spaces of our making, usually scurrying between them. Every space is an electric super-machine with the grid enlivening nearly every wall. We can take as much or little electricity as we wish. The grid is so available and so inexpensive (for all but the poorest among us) that we can easily take twice or three times as much from it on a whim.

Our electricity tastes and behaviors are not immune to some realities. We live and work and play in a wide variety of interior spaces. We're in McMansions and slum apartments, cul-de-sacs and mobile home parks, malls and big-box stores and downtown shops, hospitals and high-rise offices and factories.

All of these spaces are electric super-machines. But some prompt us to take more from the grid than others. Our grandest, immense spaces typically built today prompt us to take the most. With that much interior air to cool and heat, they inevitably order up more electricity for high-powered cooling and heating machines. Our most modest, crowded spaces typically built decades ago prompt us to take the least.

Our electricity tastes and behaviors are not immune to other realities. We live and work and play in a wide variety of ways. We're shut-in seniors or urban clubbers, working poor or country squires, farmers or farmer market shoppers.

All of these lifestyles constantly depend upon grid electricity to comfort us, convenience us, entertain us. But some prompt us to take more from the grid than others. Our most extreme, living-large, twenty-first-century lifestyles prompt us to take the most. Our most modest twentieth-century lifestyles prompt us to take the least.[32]

Humps and tails in all our distributions

You can see how diverse we Americans are. A bunch of us drink Coke and another bunch drink Pepsi. A bunch of us are raising burgeoning families in big houses. Another bunch lives alone in one-bedroom apartments. A bunch of us live in the Southwest desert or along the tropical Gulf Coast. Another bunch of us live in the snowy Rockies or along the chilly Great Lakes.

So relying upon averages when trying to understand Americans' tastes and behaviors is folly. The average of Coke and Pepsi is what? The average of big houses and small apartments is what? The average of our Southern desert/tropics and Northern mountains/lakes is what? Lukewarm mush, that's what.

Relying upon averages when trying to understand Americans' tastes in electricity and behaviors using it is equal folly. You need to probe more deeply. Averages won't cut it. You need to break down the way electricity tastes and behaviors are different and the same.

You need to break down the skewed distributions of tastes in electricity and behaviors using it. You need to find the hump on your left. That's where the majority is, at the low

32 David Owen, *The Conundrum*, Riverhead Books, 2011. "The greenest community in the United States isn't Portland, Oregon or Boulder, Colorado, but New York City... Squeezing people close together may not look green—where are their solar panels, carbon-sequestering trees, and backyard compost heaps?—but it actually reduces environmental impact, because it dramatically shrinks car ownership, makes efficient public transit possible, constrains energy use in all categories, and forces most residents to live in apartment buildings, which are among the world's most efficient residential structures... The size of the average American house has doubled since 1950, even though the average American household has shrunk. (Meanwhile, the size of the average New York City apartment, though small to begin with, has gotten smaller.) Bigger living spaces mean more raw construction materials, more furniture, more appliances, more heating, more cooling, and, in every application, more energy forever."

end of the scale. Then you need to find the tail on your right. That's where an influential minority is, at the very high end of the scale.

A majority of American households, for example, asks fairly little from our grid. They take less than thirty kilowatt-hours daily or 1,250 watts hourly.[33] At the same time, a minority of American households asks a lot more from our grid. They take more than one hundred kilowatt-hours daily or 4,167 watts hourly.

They're two totally different animals. How do you average a camel and an electric eel? Common sense tells us to try to understand the camel (the hump) and electric eel (the tail) as distinct subjects.

A minority of American households we left out are those that take between thirty and one hundred kilowatt-hours daily. No disrespect intended. The middle households, those in between the hump and the tail, are important too. They do have less influence on the distribution's measures of central tendency than the hump households and the tail households. You see, the hump households are influential because of their large numbers of households. The tail households are influential because of their enormous kilowatt-hour takes.

The middle households are just too medium. They don't have nearly as many households as the hump households. They don't have nearly as big a take of kilowatt-hours as the tail households.

That everyone is treated as one and the same by utilities and utility regulators is an expedient that serves everyone poorly. Why do utilities and their regulators do this to themselves and us?

To due or undue

"Undue discrimination" is a central principle of utility regulation. It's strictly banned, as you might have guessed. Utilities can't discriminate favoring some utility customers over others. At least they can't discriminate unduly. Utility regulators must make sure their utilities don't, according to state public service laws.

33 California households with incomes less than $20,000 paid $1.80 daily for grid electricity in 2009 on average. The median, however, is just $1.31, 27 percent lower than the average. Those who paid less than half the average, just 90 cents, were as many as 28 percent. Their homes averaged 902 square feet and included 1.9 persons. Those who paid more than double the average, $3.60, were as few as 10 percent. Their homes averaged 2,143 square feet and included 3.2 persons. The skewedness of this distribution of low-income Californians is seen generally in every distribution of household electricity usage and bills. Source: Energy Information Administration, Residential Energy Consumption Survey, 2009 microdata.

Undue discrimination is a big no. Due discrimination is okay, though. Which makes you think, what does it mean for a utility to discriminate either unduly or duly?

Everybody knows what "discriminating" means in common parlance. The discriminator treats people who are in similar circumstances dissimilarly, disadvantaging some of them.

When I was a child traveling through the Carolinas and Georgia with my family, some gas stations discriminated against African American travelers. The sign "colored bathrooms" pointed the way to inferior bathrooms.

Also when I was a child growing up in Brooklyn, New York, some house sellers discriminated against African American house buyers. Higher prices were quoted them.

Your electric utility cannot provide you inferior service or demand higher per-kilowatt-hour rates. Certainly utilities cannot and would not discriminate because of your race or other personal characteristics, dutifully obeying the civil rights laws enacted when I was a child.

Well, actually, utilities can discriminate, in certain situations, based on other factors. Discrimination in service quality or rate levels is permissible, according to state public service laws, as long as it is due. What does it take for discrimination to be due and thus permitted?

The question, it turns out, is not that easy to answer. One reason is that due or undue discrimination, either kind, is an extremely rare practice among electric utilities. So there are few precedents to look to.

Utilities naturally shy away from anything remotely looking like discrimination. Just consider that the discriminated-against utility customers might feel aggrieved, even if utility regulators pronounce that a particular discriminatory practice is due. Utilities and utility regulators generally want no part of antagonizing groups of utility customers, though there might be a rationale.

So the question about what it takes for discrimination to be due and thus permitted is usually answered theoretically, without looking to many precedents. Discrimination is due if there's a solid reason for some utility customers getting one level of service and some customers getting an inferior level. Discrimination is also due if there's a solid reason for demanding some customers pay one level of rates and demanding some customers pay a higher level.

To be discriminating

What would be solid reasons to discriminate? There are rationales that make sense. Your utility might have to incur significantly higher costs to steadily supply your electricity than for other utility customers. Discriminating against you, in this situation, might be due. That could mean inferior service for you or higher rates.

That may not seem great for you. But it's a square deal for the utility's customers as a whole. They won't be required to cover for you. Alternatively, utility regulators could see to it you get equal service and rates as everyone else. That's not fair to everyone else.

To fix your value proposition, to get you to the standard level of service and rates, utility regulators must degrade the value proposition for everyone else. The standard level of service and rates, to cover for you, must be set worse utility-wide. Lower service utility-wide than could otherwise be achieved. Higher rates utility-wide than could otherwise be achieved.

Wait! That's what most Americans have now, a sub-standard level of service and rates, all to please a small number of customers we are covering for. By not discriminating even when it is due, utility regulators are relegating citizens as a whole to a degraded value proposition for grid electricity.

Imagine the alternative. Imagine if utility regulators really followed the spirit of their state public service laws on discrimination. The regulators would order the utilities they regulate to discriminate in a few appropriate situations in which it is due, ensuring it's never undue.

The regulators would identify situations in which utilities have to incur extra costs to provide a group of customers with a steady supply of electricity. An example is a group of utility customers living in a low-density community remote from major utility facilities. Their utility operates extra substations, transformers, overhead power line structures, and power lines to supply them with electricity. When a strong storm hits the grid, the utility fields extra trucks, linemen, engineers, and construction equipment to restore the supply.

Another example is a group of utility customers living in a high-density community right near major utility facilities. The group is significantly less costly to supply with electricity than the group in the above example. And the group is significantly less costly to restore the supply when a storm hits.

Both groups of customers probably want a standard level of service, at least. To provide them that, customers in the remote low-density community should pay higher rates than those in the proximate high-density community. It would be discriminatory. But it would be due.

When regulators don't discriminate

Why don't utility regulators do this? The practice makes so much sense. But due discrimination in rates is extremely rare. Instead, a utility's residential customers pay according to a uniform rate structure throughout the entire area the utility serves. Live in a low-density community that is a two-hour drive from major utilities facilities, and your rates are no different from those living in a high-density community that is a five-minute drive from major utility facilities. It's crazy.

It's not just crazy. Shying away from due discrimination in setting utility rates is bad public policy for both groups of customers. Indeed, shying away from due discrimination is bad public policy for all utility customers.

Look at the utility customers in the high-density corner of the utility service territory. Their rates are higher than they should be to cover the customers in the low-density corner. Not only is this bad because high-density customers get higher utility bills in the mail, higher than they need be, but it's also bad because high-density customers will be less willing to go along with any utility proposals that come down the pike to raise rates. They may feel they're already paying enough.

Now look at the utility customers in the low-density corner of the utility service territory. Their rates are lower than they should be because the customers in the high-density corner are picking up part of the tab. That should make the low-density customers happy. We get lower rates than we deserve. Cool!

It's actually not so cool, and not only because somebody else is picking up part of the tab for them. If the utility is tight on funds, as every utility is these days, it looks to make cost cuts anywhere and everywhere. Guess where is a ripe place to make cuts? That's right, in the utility's expenses to support those same low-density customers so costly to serve.

If the utility cuts expenses to support the low-density customers, then that will inevitably lead to below-standard service; their steady supply of electricity may not be as steady as they would want. Their low rates, lower than if under due discrimination, come at a price. That price, inferior service, is dear.

Reddy's Recharges

1. How much electricity households take from the grid, and how much they pay for what they take, is skewed characteristically for each utility.

2. The majority of households take less from the grid than the overall average, and they pay less than the overall average.

3. Utility regulators may "due discriminate" to more fairly align the value that different utility customers get from the grid with what they pay.

ANOTHER ENERGY-SAVING IDEA FROM YOUR ELECTRIC-EEL COUNCIL

Chapter 7: What's Happening to Our Use?

Broad variation in usage and bills; climate, home size is decisive; household formation key to growth

We're atypical

What do Americans do with electricity? Our answer won't surprise readers of this book. The answer is: it varies. Electric lifestyles vary, a lot. Not only that. Their variance is asymmetric, skewed.

In this consumerist, individualist society, what Americans do with electricity is broadly distributed along several dimensions. The typical utility customer is, well, atypical.

Energy experts and utility regulators assume away the differences among utility customers with indifference. They treat the typical customer as typical.

Customizing utility service as if customers were consumers, rather than ciphers, is too complicated. It conflicts with the ideal that regulation must be rigid and regimented. It violates the unwritten rule that, rather than respond to individual differences, regulation must restrain differences.

Imagine if a business ignored the inherent variance of Americans. Imagine if a business formulated strategies oblivious to the variety in customer lifestyles. It kludges a least-common-denominator value proposition delighting no customer segment and barely satisfying the needs and wants of others. Some customer segments are turned off and take their business elsewhere. For those customers that remain, they retain little loyalty.

Utility regulators conjure up least-common-denominator value propositions. We see the same result. No utility customer segment is delighted. Other customers are barely satisfied. At least utility customers cannot take their business elsewhere. Or can they?

Incredible cacophony

It's a testament to electricity's indispensable value. What Americans do with the grid's flows is like looking through a kaleidoscope. The variety is endless as you turn the cylinder.

Life's basics—food, water, and air—are uniform when we take some. Rice is rice and string beans are string beans. Water is water. Air is air. And electricity is electricity.

The basics are all uniform when we take some. They're anything but uniform when we use some. The homogeneity of life's basics empowers us to use them heterogeneously. Water, for instance, is clean so we use it however we wish to, without giving its safety another thought.

Americans generally can count on life's basics. Being able to depend on their effortless accessibility, availability, assurance, and affordability enables us to concentrate on using rather than on taking.

Neighbors in name only

Whenever we ask, attitudes about blackouts are like a wild quilt. People's answers are seemingly uncorrelated with anything. People's region, income, house type, gender, and age don't seem to matter much. This makes no sense, at first.

One home joyfully has twins. The house is full of bustle. The home down the block of the elderly couple is a mournful quiet. The husband passed away last month. The home around the corner houses an Internet start-up in the basement. The gaggle of geeks types code through the night, pumping energy drinks.

Three detached single-family homes, all of the same size. Do they take the same amount of electricity from the grid and at the same times? Not even close. Do they use what they take in the same way? No way. Do they increase or decrease what they take at the same rate of increase or decrease? Absolutely not!

Do they have identical attitudes about blackouts? You know the answer is no. Are they equally accepting of the occasional blackout to minimize utility bills? No again.

The bustling home with the newborns can't handle any blackout of any long duration right now. Chaos would ensue. They want to know why utility regulators and utilities can't get together and make that happen.

The mournful home with the grieving widower can't handle any utility bill increase of any large amount right now. The stress would be unbearable. She wants to know why utility regulators and utilities can't get together and make that happen.

The home/Internet start-up can handle blackouts to a point and can handle utility bill increases to a point. The geeks are mobile and will plug in past a blackout's borders. The small cost of grid electricity is meaningless to them. They don't care to think about utility regulators and utilities.

It's your thing, do what you wanna do

Look first at "stationary households," utility customers who have resided in the same home for years. The data on them tells an interesting story about their electricity usage with important implications.

Some stationary households are increasing how much they take from the grid, like the bustling home. Some are decreasing how much they take. A few are increasing by a lot how much they take, like the home/Internet start-up. A few are decreasing by a lot. And many stationaries are not significantly increasing or decreasing how much they take, like the mournful home.

Considering them all together as a group, stationaries are net unchanging in how much electricity they take. Why does this happen?

Energy efficiency standards for electrical machines, appliances and devices, and efficiency retrofits may be having an effect. Take a stationary household that isn't significantly increasing how many electrical things it has and how much it's operating them. If it's recently replaced a refrigerator or lights with more efficient equipment, or if it's recently retrofitted house insulation, the take from the grid will fall a little bit. The decrease may be in the range of 5 to 10 percent.

Increased penetration of electrical machines, appliances, and devices may be having an effect too. Take a stationary household that is significantly increasing how many electrical things it has and how much it's operating them. If it's recently purchased a large TV or second refrigerator, the take from the grid will rise a little bit. The increase may be in the range of 5 to 10 percent.

A few stationary households are decreasing or increasing what they take from the grid by more than 10 percent. It's possible that's because of energy efficiency or increased penetration of electrical things. It's just as likely a household has only changed in family size.

Family size is a key factor. Electricity usage is strongly correlated with family size.

When the head of household is in his or her forties, electricity usage is generally highest. Family size peaks. Babies are born. Children haven't left yet for college, the military, or separate homes.

Younger heads of households take less from the grid. The same goes for older heads of households.

Suppose a utility's residential customers are mainly stationary households. Relatively few families are moving into the area. Relatively few are moving out. This situation is characteristic of many utilities during the deep recession that commenced in 2008.

Suppose further that births are down, elderly population is up, college population is up. This is the case nationwide. All of these factors decrease average family size and lead to less electricity usage in stationary households.

Some older children have returned to their parents' home during the deep recession. This countervailing trend increases average family size and leads to more electricity usage in stationary households.

So the data suggests stationary households as a whole are neither significantly decreasing nor increasing what they take from the grid in aggregate. Many utilities have seen near-zero growth in electricity sales. This is likely because stationary households decreasing what they take are almost exactly offsetting the stationary households increasing what they take.

This is a recent phenomenon. There was a time not long ago that most utilities saw above-zero growth in electricity sales. What happened? Stationary households contributed to this growth as penetration of central air conditioning dramatically increased. Now that central air has reached saturation nationwide, stationary households are no longer a major factor in electricity sales growth.

Moving day

Look next at "transitional households," utility customers who have changed residence in recent years. The data on them also tells an interesting story about their electricity usage with important implications as well.

Transitional households are generally increasing how much electricity they take from the grid. Few are decreasing how much they take. Some are increasing by a lot. Virtually none are decreasing by a lot. And some transitionals are not significantly increasing or decreasing.

Considering them all together as a group, transitionals are net increasing how much electricity they take. Why does this happen?

As compared to their prior home, transitional households disproportionately move into newer and larger homes. Exceptions definitely occur. Transitionals can move into older and smaller homes or those of similar vintage and size. But it is far more common that they move into newer and larger homes.

A new home probably has energy-efficient appliances and insulation. But it is likely large in home interior volume. The larger volume to cool in the summer and heat in the winter, if the heating system is electric, swamps the effect of efficient appliances and insulation.

Home interior volume, floor space times ceiling height, is a key factor. Electricity usage is strongly correlated with house interior volume. Home interior volume is strongly correlated with vintage. Newer homes are, overall, larger.

Again, many utilities have seen near-zero growth in electricity sales. As noted above, this is likely because the stationaries decreasing what they take are almost exactly offsetting the stationaries increasing what they take. And it is also likely because the number of transitional households as a percentage of all households has fallen during the deep recession. Household formation from transitionals is way down.

Transitional households drove the growth in electricity sales in the years leading up to the deep recession, in areas of rapid population growth, the Southeast and Southwest in particular. When household formation dried up in these areas, so, too, did electricity sales growth. If household formation takes off again in the years following the recession, then so, too, will electricity sales growth.[34]

Dixie power

It's not the cost of grid electricity. That's not why America has gone south. US population growth has been strong in the Southeast, the Southwest, and the South middle, and little elsewhere. But people haven't moved southward to lower their utility bills.

In fact, utility bills are substantially higher in the South. You need those higher bills there to run the air conditioners hard, to keep the interiors livable.[35]

Grid electricity is more expensive there. But it doesn't seem to matter. When people decide to uproot for Las Vegas or Phoenix or Houston or Atlanta or Miami, electricity costs don't factor in. Other considerations are far more important.

Divide the country into four broad sections, as the US Census Bureau does. The South is one of those four census divisions, as is the West, Northeast, and Midwest. In the world of the Census Bureau, the South is made up of sixteen states, from Texas, Oklahoma, and Arkansas to West Virginia, Maryland, and Delaware, plus the District of Columbia. It includes 45 million households, 37 percent of all households nationally. Its population is 112 million, 37 percent of the population nationally.

But grid electricity is especially vital for Southerners. The numbers show it. Households in the South take nearly half the electricity of households nationally. Though Southerners are

---

34 Dawn Wotapka, "Big Homes Are Back in Business," *Wall Street Journal*, June 7, 2012. "The return to bigger houses—which has taken industry watchers by surprise—indicates that the housing downturn paused, but didn't kill, America's love affair with supersize abodes… Demand for bigger houses also is coming from families that were stuck in homes that were too small but were afraid to trade up during a weak economy. Now, some say they are feeling more confident and are ready to make a change. 'We just don't have enough space,' explained … a father of three … 'The kitchen is going to be humongous compared to what we have now. We actually feel like we can live and not be bottled up.'"

35 Florida and Texas households with incomes less than $20,000 paid $3.89 and $3.69 daily, respectively, for grid electricity in 2009 on average, while California and New York households paid $1.80 and $2.52, respectively. The median daily costs for electricity were different still. In Florida and Texas, the medians were $3.64 and $3.35, respectively. In California and New York, the medians were only $1.31 and $2.05, respectively. The median in California was about one-third of what it was in Florida. So even though California and New York regulators have set some of the highest rates nationally, low-income households pay extraordinarily affordable electricity bills. And even though low-income households in Florida and Texas take much more electricity from the grid, for air conditioning particularly, their median daily costs are still in the range of $3.50. Source: Energy Information Administration, Residential Energy Consumption Survey, 2009 microdata.

37 percent of the United States, they took 48 percent of the electricity in 2011 (and paid 45 percent of the bills).[36]

The 63 percent of Americans living in the West, Northeast, and Midwest split up the remaining 52 percent of the electricity. So if you want to know what's happening nationally with electricity, look at what's going on in the swath of the country from Texas, Oklahoma, and Arkansas to West Virginia, Maryland, and Delaware.

It's a somewhat similar story for commercial and industrial businesses in the South. The sixteen Southern states took 43 percent of the electricity used by businesses nationally. Businesses in the thirty-four states of the West, Northeast, and Midwest took the remaining 57 percent.

The South didn't always dominate the country's electricity picture. In 1990, twenty-one years earlier, Southern households took 45 percent of US household electricity, and Southern businesses took 39 percent of US businesses electricity. In just two decades, the South's share has grown significantly.

36 The 42 million households in the South take 48 percent of the usage of American households and pick up 45 percent of the grid electricity costs of households, though they're just 37 percent of households. This means that only about half of electricity usage by households occurs outside of the South. Source: Energy Information Administration, Residential Energy Consumption Survey, 2009.

Graph 13:

Increasers of electricity use balance out those decreasing (one household remaining in the same home).

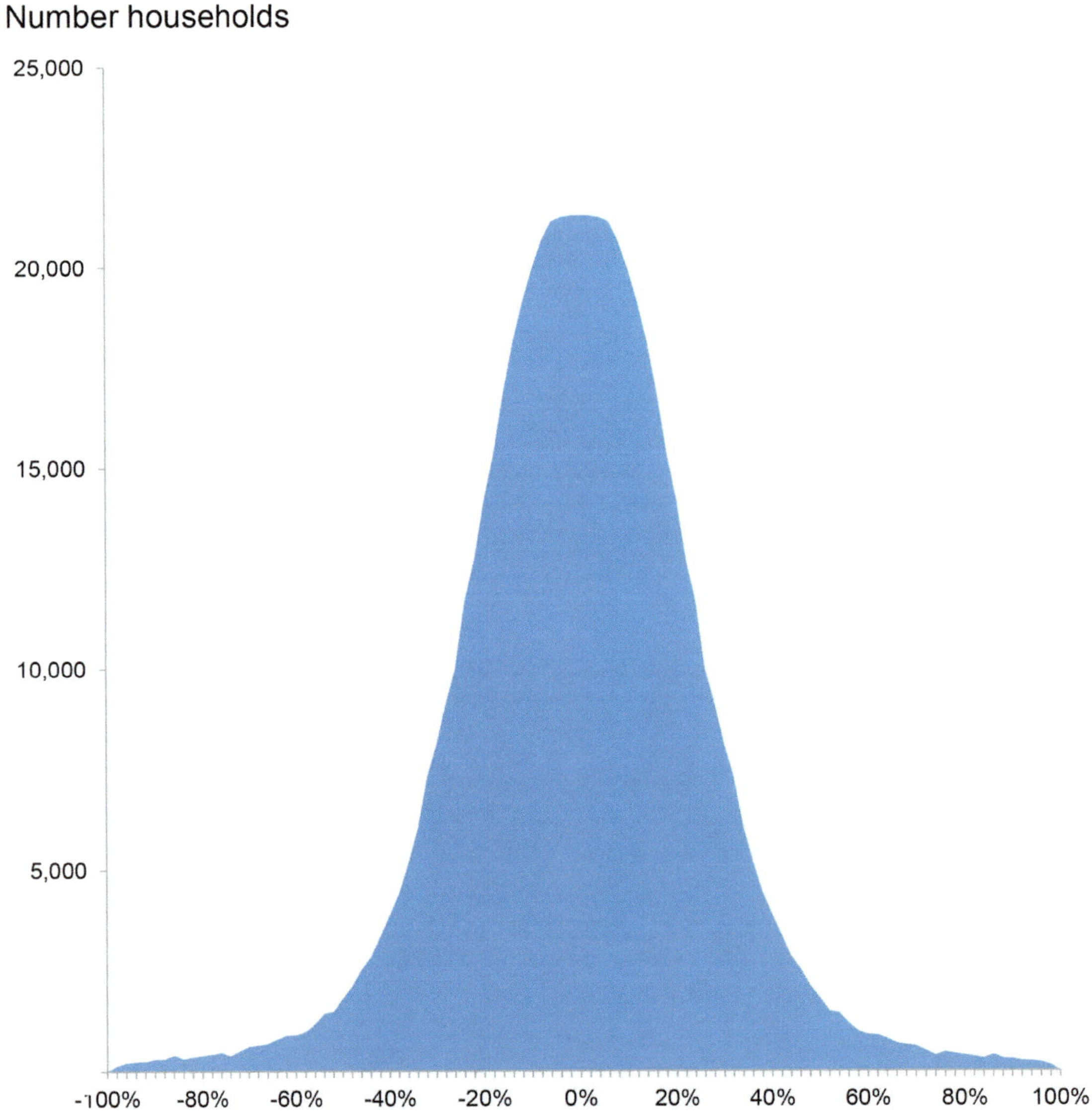

Graph 14:

Increasers of electricity use balance out those decreasing (households moving out and others in the same home).

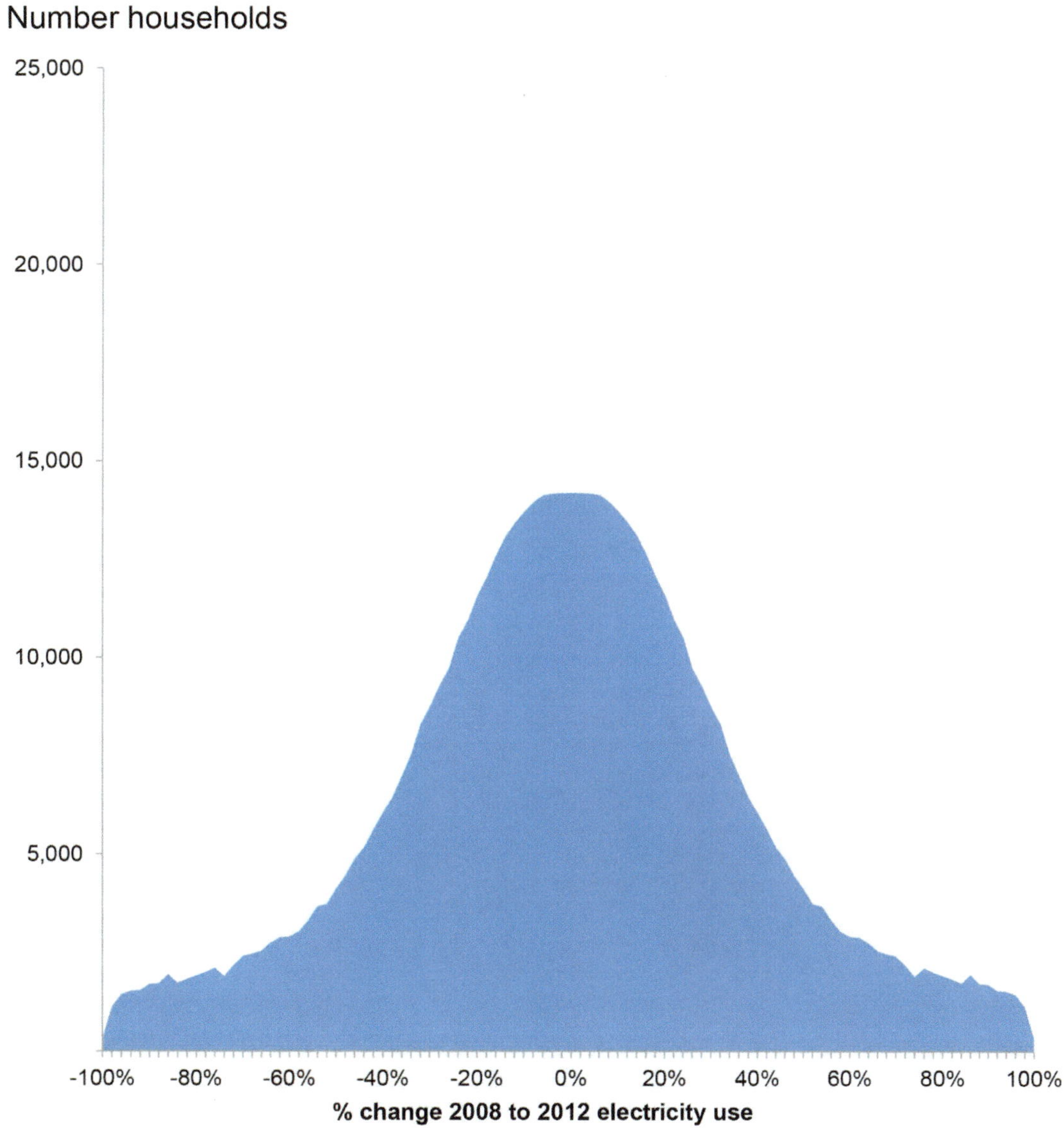

Graph 15:

Utility's electricity sales can be slightly decreasing without new homes but significantly increasing with new homes.

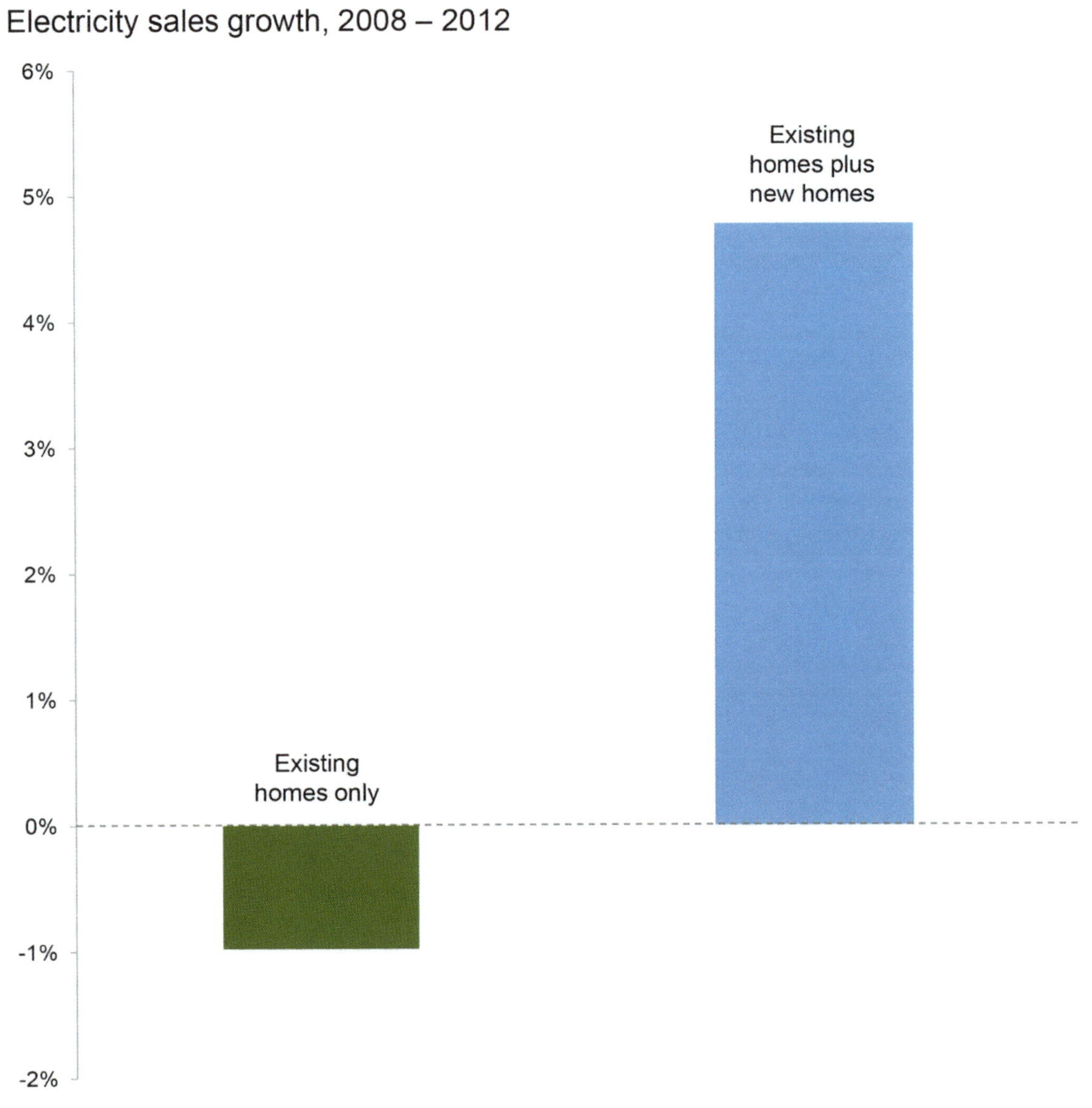

Southern households took 65 percent more grid electricity in 2011 than in 1990. In the West, Northeast, and Midwest, households took only 45 percent more. Southern businesses took 47 percent more in 2011 than in 1990. In the West, Northeast, and Midwest, businesses took only 30 percent more.

It's steamy in North and South Carolina during summertime, much less so in North and South Dakota. So naturally homes and workplaces in the Carolinas take more grid electricity to run their air conditioners. Carolinians' utility bills are consequently higher.

But utility bills in the Carolinas are higher year-round, not just in summertime. There is something else going on here, something big.

Long, hot summer

There is indeed something else going on here. And it doesn't have anything to do with the steamier Southern summers.

Southerners and Northerners alike don't intensively run air conditioners twelve months a year. Then why is electricity usage higher in the Southeast whether it's March or June, August, or November?

Southerners take more grid electricity in January. They take more in February. They take more in March, April, May, and so on through December. In comparison, Westerners, Easterners, and Midwesterners take moderate amounts, whatever the month.

Take New York City, for example. Households there only take around four hundred kilowatt-hours monthly from the grid.[37] That's an average and is smoothed out over the twelve months of the year.

How little electricity is that? It's not much more than thirteen kilowatt-hours daily. It's not much more than five hundred watt-hours hourly. We're talking about so little electricity

37 New York households with incomes of less than $20,000 paid $2.52 daily for grid electricity in 2009 on average. The median is $2.05, or 19 percent lower than the average. Those who paid less than half of the average, just $1.26, were as many as 13 percent. Their homes averaged 717 square feet and included 1.3 persons. Those who paid more than double the average, $5.04, were as few as 6 percent. Their homes averaged 2,745 square feet and included 3.5 persons. The skewedness of this distribution of low-income New York households is less than that of low-income California households in part because California regulators have set a steeply inclining rate structure. Low-usage households in California pay extraordinarily low utility bills in absolute terms and in comparison to high-usage households. Source: Energy Information Administration, Residential Energy Consumption Survey, 2009 microdata.

that it's better to express the amount in petite watt-hours rather than regular-size kilowatt-hours. And if four hundred kilowatt-hours monthly is the average, just think how many New Yorkers take less than this, a mere trickle from the grid.

It's not just New Yorkers that are spare in their electricity usage. More than 20 million households outside the South take little electricity from the grid. They're more than one-sixth of all households.[38]

All of them take less than half as much electricity as the national per-household average. Indeed, they take so little electricity, less than six hundred kilowatt-hours monthly, that energy efficiency programs cannot hope to have more than a marginal effect on their utility bills and make more than a dent on aggregate electricity usage with them.

Why is the South different? It all has to do with its unique "housing stock." In the South, housing differs in crucial ways from what we find in the West, Northeast, and Midwest.

Housing stock is the nerdy term for the mix of homes. Some homes in any area are large, and some are small. Some homes are new, and some are old. Some are single-family detached houses, while others are single-family attached houses, multi-family apartment buildings, and mobile homes.

Household formation nation

So what is it about the housing stock in the South that makes electricity usage there much higher than elsewhere in the United States? The answer is simple. It's "household formation."

As America has gone South, an enormous number of new households have formed in the region. In 2011, the 45 million Southern households were 37 percent of the national total. In 1984, the 30 million Southern households were just 31 percent. Put it this way: of the 32 million US households that were added in the last quarter-century-plus, almost half of them were added in the South. Or put it this way: one-third of all Southern households were added this recently, net.

38 Of California households with incomes of less than $20,000, 73 percent paid less than $2 daily for grid electricity in 2009 on average. Of New York households with these incomes, 47 percent paid less than $2 daily. Of Texas households with these incomes, 17 paid less than $2 daily. Source: Energy Information Administration, Residential Energy Consumption Survey, 2009 microdata.

So, well more than one-third all Southern homes are new. No other region, not the West, not the Northeast, not the Midwest, has a percentage of new homes like that.

New homes, whether in the South or anywhere, tend to have more floor space, more square footage, than older homes.[39] They tend to have taller ceilings than older homes as well. More floor space and taller ceilings translate into larger interior volumes of air to cool when it's hot out and warm when it's cold out.

New homes in any region of the country tend to be single-family detached houses, rather than single-family attached houses or multi-family apartment buildings. A greater percentage of older homes are attached and multi-family.

New homes in any region tend to be air conditioned with central air-conditioning systems rather than with window air conditioners or open windows. A greater percentage of older homes are air conditioned with windows.

Home interior volume is the key. The more interior volume of air in a home, the more electricity the home's air conditioner needs to take from the grid.

Imagine an older home that has two thousand square feet of floor space. Its ceiling height is uniformly eight feet. This means the interior volume is sixteen thousand cubic feet (two thousand square feet times eight feet).

Now imagine a new home that has five thousand square feet of floor space. Its ceiling height varies but averages ten feet. This means the interior volume is fifty thousand cubic feet (five thousand square feet times ten feet).

The new home is over three times larger. The new home has over three times the interior volume that must be centrally cooled and warmed.[40]

---

39 The 14 million households residing in homes built before 1940 paid $3.16 daily for grid electricity in 2009 on average. The 5 million households residing in homes built in the 1950s paid $3.21 daily. The 14 million households residing in homes built in the 1960s paid $3.27 daily. In contrast, the 16 million households residing in homes built during 2000–2009 paid $4.29 daily, and the households residing in homes built in the 1990s paid $4.26 daily. Older homes have significantly lower utility bills than newer homes. Source: Energy Information Administration, Residential Energy Consumption Survey, 2009.

40 The 3 million households residing in homes smaller than five hundred square feet paid $1.71 daily for grid electricity in 2009 on average. The 25 million households residing in homes between five hundred and 999 square feet paid $2.45 daily. In contrast, the 9 million households residing in homes larger than four thousand square feet paid $5.73 daily. Smaller homes have significantly lower utility bills than larger homes. Source: Energy Information Administration, Residential Energy Consumption Survey, 2009.

Pick a community, state, or region with a relatively high percentage of household formation and new homes. You can rest assured that community, state, or region will have a relatively high per-household electricity usage.[41]

Humongous

New homes with their larger interior volumes use way more electricity. The numbers don't lie. Energy experts like to think that new homes with new, efficient appliances and insulation have lower electricity usage than old homes with old, inefficient appliances and leaky walls, ceilings, and floors. They don't. It matters much more that new homes' interior volumes are, in general, jumbo-sized.[42]

The implications are humongous too. Suppose you're a utility serving an area with robust household formation. Individuals and families are moving in, in numbers. Their homes are disproportionately larger in interior volume. So they're driving up the percentage of utility customers that are heavy-usage households. Your sales of electricity will grow at a hefty pace.

Suppose instead you're a utility serving an area with negligible household formation. Individuals and families aren't moving in, in any appreciable numbers. So the percentage of utility customers that are heavy-usage households is static. Your sales of electricity will grow at a slow pace or no pace at all.

Readers can't get through a chapter of this book without being reminded to look at distributions rather than averages that mislead. When the US Department of Energy reports that sales of electricity are barely growing nationally, know that sales are growing robustly in some areas and falling in other areas.

---

41 Joel Kotkin, "America's Growth Corridors: The Key to National Revival," Manhattan Institute for Policy Research, 2013. "Future American economic growth will not be centered in the denser regions of the East and West Coasts of the United States that are often lionized in the national press. Instead, growth in the near future (and likely beyond) will be found in four broad regions that are already doing well, despite a devastating recession and weak recovery. Even as many Americans lower their expectations of future prosperity for themselves and their children, these regions look forward to robust expansion in the years ahead... Migration from abroad has followed the same general pattern as migration within the U.S.: the growth corridors are attracting more new arrivals than ever... Over time, we expect that overall migration patterns driven by housing costs will shape educated migration, too. This trend could become even more pronounced when the housing market recovers further and homeowners in the Northeast are again able to sell their houses."

42 Lam Thuy Vo, "American Houses Are Still Much, Much Bigger than They Used to Be," Planet Money Blog, Aug. 23, 2012. "Newly built homes got a little smaller during the housing bust. But just a little. As of 2010, the median home was still bigger than homes built in the 1990s, when the economy was booming. And the median home built in 2010 was much, much bigger—41 percent bigger—than the median home built in 1975."

Electricity sales are growing robustly in areas of strong household formation. Sales aren't growing, or are slightly shrinking, in areas of weak or insignificant household formation.

In the deep recession where household formation has been weak or insignificant nationally, electricity sales are not growing or slightly shrinking nationally. When the United States exits the deep recession, household formation will once again be strong in some areas, and those areas will resume robust electricity sales.

The disparity in electricity sales growths, from utility to utility, is cavernous. A utility's electricity sales growth is highly correlated with household formation and its proxy, the utility's customer count growth. Housing formation sets in stone a utility's destiny.

Deciding your utility bill

It's come time for the big decision. It's come time to move. The current place just won't do anymore. It's too small or too large. It's either too far from work or near to work but too expensive. It's in too busy an area or an area with not enough going on.

A new job or a new relationship, or a newly broken relationship, makes the need to move a must. A new addition to the family, a birth, cries out for a larger place. A loss, a death, cries out for a smaller place. The kid moves out to college or a job, so the old house is too empty a nest. Or the kid moves back in, so the old house is crowded as never before.

Where you choose to move sets in motion what you'll take from the grid. Automatically, if you move South, you'll take more electricity. Automatically, if you move into a single-family detached house, you'll take more electricity.

Tell me where you're moving to, what region and what kind of home. I'll tell you your level of electricity usage with amazing accuracy. Give me some more. Tell me the home's square footage or the number of bedrooms. I'll tell you your level of usage with even better accuracy. Where you move to is totally telling, in terms of electricity usage anyway.

An immigrant moves to an apartment on the island of Manhattan. US electricity sales increase by only thirty-five hundred kilowatt-hours per annum. What if that same immigrant instead moves to a suburban house around Houston? US electricity sales increase by five times as much. What a big difference, all because of where our immigrant moves.

For every five immigrants moving to a Northern city apartment, one immigrant moving to a Southern suburban house has as much impact on US electricity sales growth. A similar effect occurs whenever anyone moves from a Northern city apartment to a Southern suburban house. US electricity consumption growth is dictated by household formation that is disproportionately Southern and suburban.

It is where that new household is added. That's the determinant of what happens to our overall electricity take. A household's behavior toward electricity usage and conservation has a marginal effect.

The most profligate apartments in Manhattan can't crack one thousand kilowatt-hours monthly. Movie stars, sports heroes, Wall Street magnates all living it up in their luxurious addresses guarded by broad-shouldered doormen can't use more kilowatt-hours than that no matter how hard they try.

It's true, no matter what they're doing up there in their glamorous suites. The interior volumes are limited. The outdoor climate is relatively mild. The apartments are densely packed within their stately buildings. All of these factors put a hard cap on how much electricity they take from the grid.

The most energy-efficient houses around Houston can't fall short of one thousand kilowatt-hours monthly. Even the greenest families eagle-eyeing their utility meters and installing insulation and compact-florescent bulbs all over can't use less kilowatt-hours than that no matter how hard they try.[43]

It's true no matter what they're doing over there in their Spartan houses. The interior volumes are huge. The outdoor climate is relatively harsh. The houses are widely separated from neighbors by crew-cut lawns. All of these factors put a hard floor on how much electricity they take from the grid.

Are our new immigrants moving into apartments in Manhattan or Boston or Detroit or Denver or San Francisco? Or are they moving into houses around Houston or Atlanta or Miami or Phoenix or Las Vegas? When immigration is robust, the new households disproportionately form in Southern suburban houses. US electricity consumption grows robustly too. When immigration screeches to a halt, US electricity sales growth screeches as well.

---

43 Stan Cox, *Losing Our Cool*, The New Press, 2010. "The authors [of a 2005 study published in the *Journal of Industrial Ecology*] concluded that 'a 1,500-square-foot house with mediocre energy performance standards will use *far less* energy for heating and cooling than a 3,000-square-foot house of comparable geometry with much better energy detailing.'...The big-house trend pushes up [energy] consumption... Despite consuming far more total energy, the owner of a big, green house can boast of much bigger energy savings than the small-house owner." *Emphasis added*

Smart boring meter

Why hasn't the offsetting of stationary households been noted by energy experts? The tendency of increasing usage households to cancel out decreasing usage households is too important to be missed.

One big reason is that electricity usage is very highly volatile. Look at your own household. How much electricity do you use minute to minute, hour to hour, day to day, month to month, year to year? Few have the means and inclination to check. But, be assured, your usage varies constantly, crazily

Just take a look at your new smart meter, if your utility has installed one for you. I did this for my own house in Washington, DC, a couple of years. My utility hadn't yet installed a smart meter. I couldn't wait. Searching on the Internet, I found a quick fix. The cool device arrived about a week later. So simple to install, even I could hook it up onto the front of my un-smart meter to intercept its secrets.

The device's cousin was also in the box. I gave it a seat of honor on the kitchen table. This second part was in constant contact with the hookup to the meter. Now we could tell how many kilowatt-hours we were taking from the grid in real time, at any instant in time.

My wife, daughter, and I couldn't keep our eyes off of it, at first. One moment the house was taking four kilowatts. The next the house was taking one kilowatt. It was fascinating and a topic of conversation in our family but for just a few days. After that, it was a total bore.

After a few days, we could shut our eyes and guess fairly well what the device was saying, how much electricity the house was taking. When the air conditioner cycled on, obvious from its muted roar, you knew the house would be taking at least three kilowatts. When the kitchen water cooler cycled on, you knew the house's take would bump up a half kilowatt. If you were wandering through the kitchen at night to secret a snack, you knew the house's take would be less than a kilowatt-hour. It was all so predictable and boring.

Declining and inclining

Earlier we assumed utilities charge a flat per-kilowatt-hour rate. For every kilowatt-hour you take from the grid, your utility will bill you about a dime.

What if your utility instead offers a quantity discount? Take one thousand kilowatt-hours in a month or take any lesser quantity. Your utility will bill you about a dime apiece. But...

What if you take more than one thousand kilowatt-hours? Suppose you take fifteen hundred kilowatt-hours. Your utility bills you about a dime for each of the first thousand kilowatt-hours, as before. But it bills you just eight cents for each of the extra five hundred kilowatt-hours. That's a two-penny discount. It translates into ten dollars per month off. You can almost treat yourself to a movie.

What if your utility instead has a quantity penalty? Again you take fifteen hundred kilowatt-hours. Your utility bills you about a dime for each of the first thousand kilowatt-hours, as before. But it bills you twelve cents for each of the extra five hundred kilowatt-hours. That's a two-penny premium. It translates into ten dollars per month additional. Forget the movie. Watch cable at home.

In energy circles, we call a utility pricing structure with a quantity discount a "declining block structure." Each block of greater kilowatt-hour usage costs a declining amount in dollars and cents. We call a utility pricing structure with a quantity penalty an "inclining block structure." Each block of greater kilowatt-hour usage costs an inclining amount in dollars and cents.

Heavyweights

Heavy-usage households in California get socked in two ways. They pay more for grid electricity than other households because they take more of it. That seems fair. They also pay more because their per-kilowatt-hour rate is higher under California's inclining block structure. Does that seem fair?

Let's look at the inclining block structure for California Dreamin' Power Company. Its per-kilowatt-hour rate for most households, whether they are light users or medium users, is twelve cents. Their usage falls entirely within this inexpensive initial block that covers whatever households take from the grid up to fifteen hundred kilowatt-hours monthly.

But for any household that takes more than fifteen hundred kilowatt-hours monthly, an expensive inclined block kicks in. For example, if a heavy user takes three thousand kilowatt-hours in a month, the first fifteen hundred kilowatt-hours is billed at twelve cents per and the second fifteen hundred kilowatt-hours is billed at thirty cents per. There's a steep price to pay for blowing past the inexpensive initial block and getting socked by the expensive inclined block.

Suppose 10 percent of California Dreamin' residential customers are heavy-usage. Suppose further they each take three thousand kilowatt-hours in a month.

The other 90 percent of the customers are not heavy. They're equally divided, we'll say, between light users and medium users. They each take less than fifteen hundred kilowatt-hours, safely within the boundaries of the inexpensive initial block. We'll assume each light-usage household takes six hundred kilowatt-hours, and each medium-usage household takes fourteen hundred kilowatt-hours.

Doing the math, the heavy-usage households cover as much as 31 percent of the utility bills paid by the California Dreamin' residential customers though they're just 10 percent of the customers. Heavy users make it easier on some of their neighbors.

The medium-usage households cover 48 percent of the bills. This is roughly in line because they're 45 percent of the customers.

However, the light-usage households really get spared, thanks to the heavy users. Light users cover only 21 percent of the bills though they're 45 percent of the customers.

The same goes when utility regulators set a utility rate increase. The heavy-usage households cover as much as 31 percent of the rate increase. All other households, light users and medium users, cover just 69 percent of the increase.

Consider when California Dreamin' proposes a rate increase of $300 million dollars. A proposal of that magnitude is a guaranteed headline-grabber.

Let's simplify and assume California Dreamin' doesn't have any business customers to spread some of the rate increase onto. It only has the residential customers we've been discussing.

The 10 percent of customers who are heavy users would cover 31 percent of the increase, or $91 million. The 45 percent of the customers who are medium users would cover 48 percent of the increase, or $144 million. But the 45 percent of customers who are light users would cover only 21 percent of the increase, just $63 million.

If California Dreamin' has 3 million residential customers, the 45 percent that are light users number 1.35 million. In the event California utility regulators approve the rate increase, the $63 million annually, $5.25 million monthly, would be spread among all those light users. The impact on each of them would be $3.89 monthly or thirteen cents daily, an increase of 5 percent. Because of the heavy users, the light users get by with a fairly light increase in their grid electricity costs.

Affluent influence

That 30-percent minority of customers foots two-thirds of the dollars. The 70-percent majority need only foot the remaining one-third of the dollars. Lucky for them! This is true whether we're talking about the utility's total billings to residential customers or about the total increase in billings when utility regulators set rates higher.

The 30-percent minority is made up of medium-usage households and heavy-usage households. These are generally the more affluent households served by a utility. By paying the lion's share of a utility's billings to residential customers and the lion's share of the increase in billings when regulators set rates higher, the more affluent hold down what the less affluent have to pay.

So a utility's costs to keep its grid humming and the increases in costs it must bear are progressively spread among customers. The more affluent pick up the slack for the less affluent.

A utility's costs are so progressively spread that the percentage of households' expenditures taken up by electric bills is remarkably level across income levels. Nationally, the 122 million American households paid 2.9 percent of their total expenditures on electric bills, on average. How did this percentage vary by income?

The less affluent did pay a higher percentage than the 2.9 percent, but not much more. The 8 million households with 2011 incomes of ten thousand dollars to $14,999 paid the highest percentage of any income group. Of their household budgets, 5 percent went to pay their electric bills. For the 8 million with incomes of fifteen thousand dollars to $19,999, 4.5 percent went to electric bills. For the 5 million households with incomes of five thousand dollars to $9,999, 4.3 percent went to electric bills.

These are among the poorest Americans. The financial strain on this 17 percent of all households is significant. That being said, electric bills accounted for just 4.3 percent to 5 percent of their household budgets.

No other income group spent more than 4 percent of their household budgets on grid electricity, on average. This includes the 5 million households with incomes of zero to $4,999. Though the average for all households is 2.9 percent, very few households pay a significantly larger chunk of their budgets. Those that do, even the $10,000 to $14,999 income group, aren't that much higher.

Notably, those with incomes of $10,000 to $14,999 have the lowest total expenditures of any income group, even lower than those with incomes of zero to $4,999 and those with incomes of $5,000 to $9,999. Those with incomes of $10,000 to $14,999 also have the fewest persons and fewest earners within their households. They are the oldest in age and the highest in female-to-male ratio. They receive the largest share of their income from Social Security and other retirement sources of any income group.

Reddy's Recharges

1. American households vary broadly in their usage of grid electricity, but households in new homes generally have above-average usage.

2. Stationary households with increasing and decreasing electricity usage offset each other, leaving new households determining overall electricity sales growth.

3. Generally affluent heavy-usage households cover a large proportion of utility bills and rate increases so less-affluent light-usage households can pay a small proportion.

"Is it dark in here or is it just California?"

How We *Value* Grid Electricity

On why utility regulation has been successful, why it recently has lost focus on the value of grid electricity, and how it can refocus

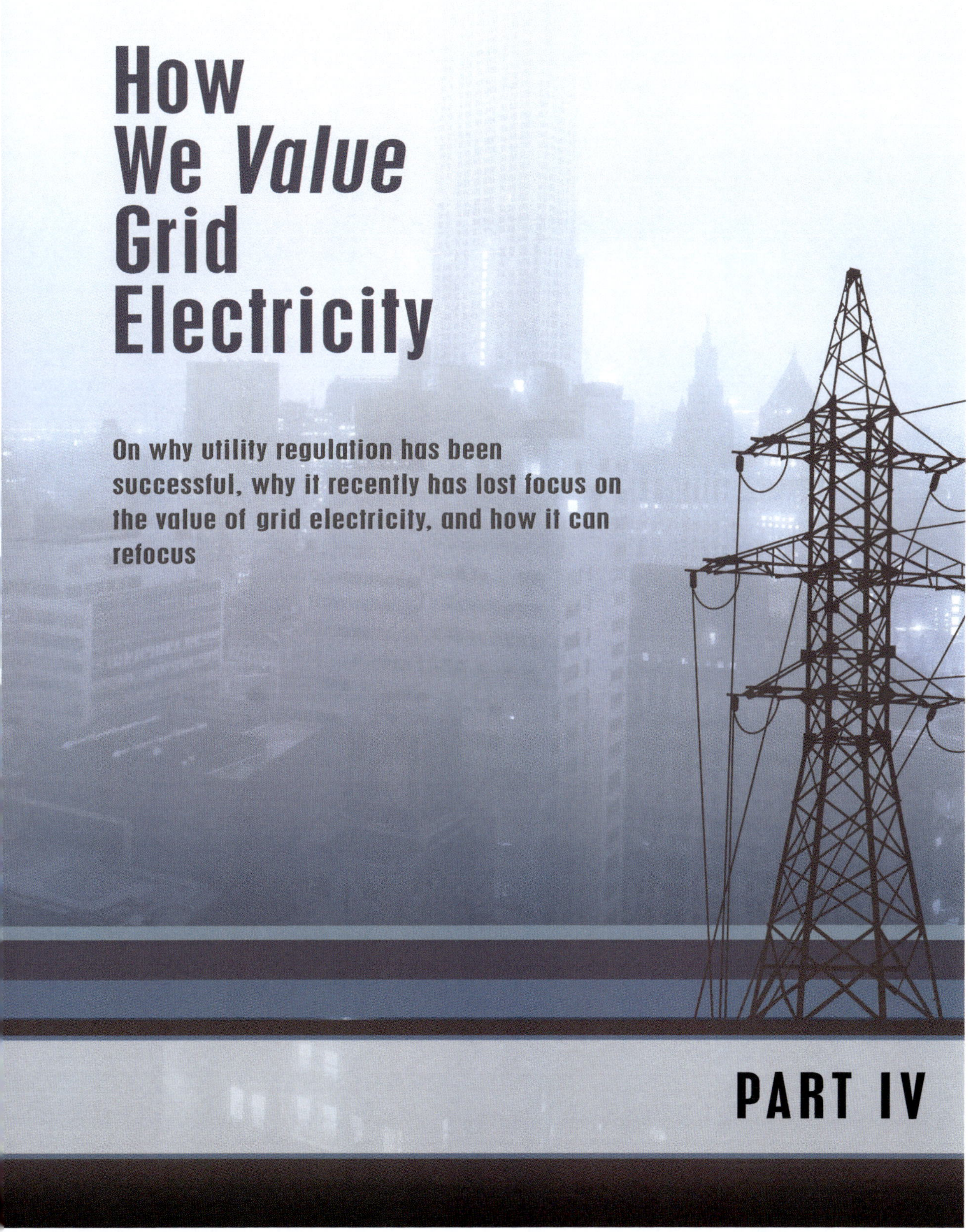

PART IV

Part IV. How We *Value* Grid Electricity

Chapter 8: Regulation's Focus on Value Then and Now

Chapter 9: Enter Lengthy Storm Outages

Chapter 10: Refocusing on Value for Money

Without limit

In the twentieth century, we grid-enabled every building in the country, no matter how remote, no matter the economic circumstances of its owner. We lighted every building and made their interior climates temperate, day as well as night, spring and fall as well as summer.

We cooled and froze food and beverages everywhere. Now we eat and drink any food and beverage without respect to season, without respect to where and when it was grown and where and when it was prepared.

The grid gave us the power to pump, grind, mix, cut, slice, sharpen, sew, wash, dry, press, vacuum, brush, cook, toast, and warm. With it, we print, copy, count, calculate, listen, speak, amplify, amuse, and play. We time, monitor, alert, protect, build, sort, package, promote, sell, buy, magnify, display, teach, study, write, read, comfort, and cure.

The grid does all this. It does these things ubiquitously, inexpensively, invisibly, and without limit. So we use the grid's goodness at every opportunity to better work, relax, rest, learn, create, communicate, and care. Of course this greatly stimulated productivity and thereby transformed the economy.

Decelerating

There has been a long-term deceleration in grid electricity sales. There's no disputing it. The trend is largely the product of three factors. First, central air conditioning reached saturation. Second, manufacturing fell. Third, the electricity needed to run the latest models of machines, appliances, and devices fell.

Central air has been installed in well over 100 million residential and business buildings. This has been enormously transformational and beneficial. It made it possible to further populate the country's Southern tier, where exterior temperatures are often high. It catapulted productivity, on hot days particularly. It prolonged people's lives.

By this time, there are fairly few buildings that don't have central air. They tend to be in the country's Northern tier, where exterior temperatures are rarely high. Their interior volumes—to be cooled—tend to be small. So when central air is installed in the buildings still on room air conditioning, or without any air conditioning, the incremental impact on grid electricity sales is modest.

During the last decades of the twentieth century, the driving up of grid electricity sales by central air installations was increasingly counteracted by deindustrialization. Now that installations are driving up sales by much less, here in the twenty-first century, the combination of this and deindustrialization is producing little growth.

We have packed many more machines, appliances, and devices into buildings. This should have spurred grid electricity sales. However, we've substituted many old machines, appliances, and devices with new ones that do their thing with fewer watts. The net effect is nil. More electrical equipment, more useful equipment, but roughly the same take from the grid.

Formation

Is grid growth and economic growth delinked? Critics of the grid clearly relish saying so. We can agree that grid growth in existing residential and business buildings is now pretty flat. Grid growth and economic growth isn't coming from existing buildings.

But are we forgetting something? Isn't there another key component of the US economy?

Sure there is. It's new buildings. It's housing formation from new residential buildings and business formation from new business buildings.

Every new building—residential or business—disproportionately adds to grid electricity sales. One new building can take as much electricity from the grid as the net-change in the electricity take of a hundred existing buildings or maybe more. So when we slow down the construction of new buildings because of an economic growth slowdown or a population growth slowdown, we also slow down the growth of grid electricity sales.

When housing formation and business formation were robust in the mid-2000s, grid electricity sales grew vigorously with the economy. When housing formation and business formation collapsed in 2008–2009, grid electricity sales fell with the economy.

So there's been no delinking, despite how in vogue the delinking theory is. What we've seen since 2008–2009 is the linked fall of business formation, housing formation, and grid growth all in tandem.

When business and housing formation picks up again, where it most vigorously picks up, grid electricity sales shall pick up too.[44] Grid growth and economic growth will be strongest in these regions of the country.

Grid sales, grid value

We miss the bigger picture when we fixate on grid electricity sales. It's understandable why some trumpet the grid sales trajectory. Those with agendas to shrink grid sales—businesses competing against the grid, backed by interest groups contending with the grid—gleefully point to the decelerating trend. For them, it's welcome proof of what they yearn for.

The scenario they sell is an America that, while not grid-free, is grid-light. Grid sales will decelerate then decrease then decrease dramatically in favor of competitors like home-brew electricity and the latest fad, community-brew electricity ("micro-grids").

It'll be great, they promise. There'll be strong growth for the grid's competitor business models. And, yes, there'll be stunted growth for that stodgy old grid model. This, from their perspective, is just perfect.

So, naturally, the marketing machine for the grid's competitors wants us to fixate on grid sales growth. The grid is going down. Can't you see? We—the burgeoning solar-panel and gas-generator industries, demand-response financial traders, grid-infrastructure

44 Rebecca Smith, "U.S. Electricity Use on Wane," *Wall Street Journal*, Jan. 2, 2013. Despite the headline, the article reports: "The Energy Information Agency is projecting that electricity use in the U.S. will *rise* an average of *just* 0.6% a year for industrial users and 0.7% for households through 2040." The article didn't note that the EIA projection for growth in electricity use by commercial users is higher, 0.8% a year. *Emphasis added.* Source: EIA AEO2013 Early Release Overview, Dec. 5, 2012.

adversaries, energy efficiency grant recipients, researchers, consultants, lobbyists, academics, equity funds, et al.—are the future.

Let the grid's competitors have their say. Their sales growth is indeed phenomenal, thanks to preferential government policies and utility regulation. Energy efficiency grantees, for example, will soon be grossing $20 billion annually. Nice. But that doesn't mean we have to buy into the transparent pitch of the competitors.

Growth that counts

Grid growth in terms of grid electricity sales has been decelerating for decades. Okay. But grid growth in terms of the grid's importance to America and value to Americans is accelerating, in the last couple of decades particularly. Let's fix on that trend for a minute.

In 1952, the year I was born, US households and businesses took around 0.35 billion megawatt-hours from the grid. In 2012, when I was sixty, we took around 4.10 billion.

So we took 11.7 times the electricity in 2012 than in 1952. That's nearly a twelve-fold increase during my lifetime. That's a lot.

Most of the increase happened during the first three-quarters of the sixty-year period, during 1952–1997. In 1997, when I was forty-five, we took around 3.50 billion megawatt-hours from the grid.

So we took ten times the electricity in 1997 than in 1952, a 900 percent increase. And we took just 1.17 times the electricity in 2012 than in 1997, a 17 percent increase. Clearly, grid electricity sales growth has slowed.

But wait, population growth also slowed. From 1952 to 1997, during my first forty-five years, the nation's population increased 1.7 times, 70 percent, from 158 million to 268 million. From 1997 to 2012, during my next fifteen years, the population increased 1.17 times, 17 percent, from 268 million to 313 million.

Did you spot the coincidence? From 1997 to 2012, grid electricity sales increased 17 percent, as did the population. So per capita electricity sales—sales divided by population—remained constant.

From 1952 to 1997, however, grid electricity sales increased 900 percent, while the population increased 70 percent. So, per capita electricity sales didn't remain constant during these years. What we took from the grid per person increased by 490 percent.

Per person grid electricity grew wildly during my first forty-five years and then not at all in my next fifteen years. Why? What happened?

Sales growth then value growth

It was a revolution. Tens of millions of buildings standing in 1952 loaded up on electrical machines, appliances, and devices over the next forty-five years. Tens of millions got window air conditioning. Window air conditioners cooled interior spaces but also upped the kilowatt-hour take from the grid. Then these were pulled out in favor of central air conditioning. Central air conditioners further cooled interior spaces, prettied up the windows, and further upped the kilowatt-hour take.

Tens of millions of buildings were put up during the forty-five years. Each came grid-enabled and quickly loaded up on machines, appliances, and devices. But they never stopped.[45] The revolution swept them too. Some came equipped with window air conditioners. But soon they upgraded as well to central air.

When there was a blackout in 1952, households and businesses were inconvenienced to a minor extent. The only machines, appliances, and devices in most households then were a few lights, a refrigerator (with a miniscule "deep freeze" space), a sewing machine, clothing iron, vacuum cleaner, toaster, fan, phonograph, and radio. In a power outage, you could easily get by without them.

When there was a blackout in 1997, households and businesses were much more inconvenienced than in 1952, forty-five years earlier. In 1997, households commonly had air conditioners, televisions, a washing machine and dryer, a video recorder, even a computer. But few depended upon the Internet. And no one had smart phones. So blackouts were still palatable, however unpleasant.

Telephone service, with its own independent power source, was still the main mode of communications. Books, magazines, and newspapers were still how many of us kept up

45 In the last four years alone, January 2009–January 2013, US household penetration of high-definition TV increased from 50% to 88%, laptops from 50% to 71%, tablets from zero to 44%, home networks from 34% to 61%, and smart phones from 23% to 55%. Source: Consumer Electronics Association, *U.S. Consumer Electronics Sales and Forecasts 2008–2013*, Jan. 2013.

with events, learned, and passed time. Retailers and banks could still serve customers with people, pens, paper, and cash registers.

When there was a blackout in 2012, households and businesses were devastated, and some were left helpless. Communications and commerce were crippled. So the cost to utility customers of grid blackouts has soared. This means that the value to utility customers of grid electricity has also soared. Because of the last decade's Internet transformation, the grid's value has shot up far faster than the 17-percent increase in the grid's sales.

Segway spin

It wasn't long ago. People traveled sidewalks by walking mostly, or with a wheelchair, or on a bicycle or skateboard. Have you noticed that something has changed? A few people are now traveling sidewalks on electricity.

On the electric-powered Segway, pedestrians can go further faster. Law enforcement naturally loves them, extending the range and frequency of foot patrols. On the electric-powered wheelchair, the handicapped and elderly can get around. These are wonderful inventions. Who can imagine all of the future applications? Will it spark a renaissance of downtown and main street shopping? How long will it be before millions of Americans power along sidewalks?

Not long ago, people traveled between floors of their homes by walking up and down stairs. Have you noticed that something has changed? A few people are now traveling between floors on electricity.

On the electric-powered home elevator/lift, the handicapped and elderly can better enjoy their home life. How long will it be before millions of Americans power up and down in their homes?

How long will it be before millions of Americans power-cool themselves outside during hot summer days? How long before they power-warm themselves outside during cold winter nights? Will it spark a renewed interest in spending more time outside? How long before Americans magnify their sight and amplify their hearing as if electrified supermen and superwomen? Will this increase safety during nighttime?

How long before we noise-cancel the bothersome sounds outside the home? How long before we use voice-command artificial intelligence agents to operate machines, appliances, and devices inside the home? Actually, that revolution has already started.[46]

You get the point. The constant flow of new electrical wonders isn't slowing anytime soon. Who thinks Americans are going to use the same number and variety of powered things in 2040 as we do today?

Not even Steve Jobs was prescient enough. Not even he could envision how American life will be transformed, decades out, by technological innovation. Will we be stronger? Will we be safer? Will we be smarter?

It wasn't long ago. People couldn't have a conversation with faraway family or friends while walking or driving. People couldn't send a note to them that same day. They couldn't watch any movie or read any book whenever, wherever. They couldn't get an MRI at the shopping mall. They couldn't go into a store without cash or make a purchase without a store.

Innovation with electricity hasn't peaked. That's silly. It's still going strong. In fact, we're just getting started.[47]

---

46 Carla Diana, "Talking, Walking Objects," *New York Times*, Jan. 27, 2013. "The behaviors of these future robotic objects may be utilitarian, like a lamp that bends to follow the items you reposition on a desk, or they may provide feedback, like a fork that vibrates when you're eating too fast… A robotic cutting board might guide you while you cook and offer helpful tips when your knife skills aren't up to par… A raincoat might glow or whistle when it knows you'll need it on your commute… A medicine bottle might open its lid to offer you a precise dose or automatically call your pharmacist when it's empty… The future will be rich with sensor-based, animated objects using expressive sound, light, motion and screens to praise, encourage, advise and comfort us. We'll have fun with TVs and music players that reposition themselves in response to our dance moves, we'll be relieved of chores by robots that pick up our crumbs and scrub the floors, and we'll stay fit with robotic parasites that curl around our wrists to take health stats and exhort us to get off the couch."

47 Robert J. Gordon, "Is U.S. Economic Growth Over? Faltering Innovation Confronts the Six Headwinds," National Bureau of Economic Research, Aug. 2012. "A common feature of this innovative revolution was that many of the improvements could only happen once. Speed of travel was increased from that of a horse to the jet plane in a century but that could not happen again. The interior temperature that in 1870 alternated between freezing cold in the winter and stifling heat in the summer reached a year-round 72 degrees Fahrenheit, and that could not happen again. The U.S. was transformed from 75 percent rural to 80 percent urban, and that could not happen again… Attention in the past decade has focused not on labor-saving innovation, but rather on a succession of entertainment and communication devices that do the same things as we could do before, but now in smaller and more convenient packages." This author—among many—is critical of Gordon's pessimistic conclusion that innovation has nearly come to an end. The pessimism is shared by those who expect that coming decades will see few new compelling electrical machines, appliances, and devices. As this book suggests, the Age of Electricity is still in its prime. Innovators of the next generation will create unimaginable "apps" to provide comfort, convenience, and entertainment that consumers will crave and that will further increase human collaboration and productivity. As just one example, Gordon downplays the potential for the Internet as a platform for innovation. If Gordon were writing in 1910, he might have downplayed the potential for the already-invented automobile, airplane, and the electricity grid as platforms to spur collaboration and productivity in the next hundred years.

It makes you laugh at the experts' confident forecasts of future electricity use. The experts act as if they know all of the ways electricity will be used ten, twenty, thirty, forty years out. They act as if they can extrapolate future use from current use.

Predicting electricity use is like predicting future Internet use. Both the electricity grid and Internet grid are organic innovation platforms not products. The cleverest futurist can only guess how much and how they will grow.

Now for Part IV

In Chapter 8, "Regulation's Focus on Value Then and Now," you'll read about how the team started to tell and testify about our discoveries, tentatively at first, more confidently thereafter. It was exciting how our findings clicked with so many people. This stirred us on.

In Chapter 9, "Enter Lengthy Storm Outages," you'll see how we stumbled upon our fourth mega-discovery in a year's time. After finding that electricity costs are surprisingly moderate for most utility customers (despite the contrary common view), that the statistical distributions of electricity usage and costs are skewed, and that electricity sales growth is driven by housing formation and economic formation, we thought we were done. No, we weren't.

Asked to look at how utilities tackle power outages, we found that the statistical distributions of blackouts are skewed too. And that recognizing the skewedness in blackout duration and outage costs transforms your perspective on how to best address the problem.

In Chapter 10, "Refocusing on Value for Money," you'll learn how the team tied all of the new principles of electricitynomics together in a "unified theory." Our fifth mega-discovery was that the economics of electricity usage and bill costs and the economics of outage duration and blackout costs are inextricably linked. The crucial link is utility customer value.

In a nutshell, a customer's bill costs are well less than his or her blackout costs. This means utilities, in partnership with utility regulators, are likely able to reduce customers' blackout costs significantly more than they increase customers' bill costs, thereby increasing customers' value.

The journey began innocently enough with an itching curiosity about electric rates and bills. Part I told the story of these first tentative steps. Soon after, the journey was transformed into a wondrous tour, full of surprises. The team had its first inklings about electricity's value. Part II told the story of these surer steps. Part III recounted the growth spurt, when we experienced a surge of discoveries, once we were able to lay our hands and computers on hundreds of millions of utility records.

Part IV will show us a path forward. As you shall read, utilities and regulators now have the analytic tools to measure the value that the public nets from the grid and the tools to maximize that value.

Households in large, single-family detached homes typically have higher incomes and take more electricity from the grid than households in small, multi-family apartment buildings.

Low-income households therefore tend to pay little for electricity.

Chapter 8: Regulation's Focus on Value Then and Now

Regulation founded to accelerate investment; focus elsewhere now; essential costs; allocation; value introduced

Teddy, William Howard, and Woodrow

Regulation of electric utilities was literally invented over eighty years ago in the roaring 1920s and depressing 1930s. States wanted to accelerate utilities' capital investment in the grid. They would grant monopoly franchises to utilities with exclusive service territories. With guaranteed markets of consumers clamoring for electricity, states knew the franchised utilities would be able to finance the millions needed. (Millions was a lot of money then.)

The public craved electricity's first apps, even throughout the Great Depression. Lighting was just the start, like e-mail was for the Internet. Then, in quick succession, there were phonographs and radios to entertain and inform, irons and vacuum cleaners to tend house, refrigerators and fans to cool our food and us, pumps and motors for factories, and elevators for tall buildings. Nearly every year, more cool apps of grid electricity premiered.

But a big worry accompanied the franchising of utilities. States feared the greedy impulses of utility owners. There was a real concern about this in the 1920s and 1930s. A sinister utility owner, left unconstrained, could turn rapacious. After all, his franchise licensed him to operate freely as a monopolist in a section of the state, without a care about competitors.

Everyone in the period recalled the pitched battles of the administrations of Presidents Theodore Roosevelt, William Howard Taft, and Woodrow Wilson. These were epic clashes to curb monopolistic behaviors seemingly running rampant throughout the economy.

So states packaged in strict oversight. The overseers, permanent commissions of regulators, were empowered to probe and punish monopolistic behaviors.[48]

The franchising and regulating of utilities, packaged together, achieved incredible success. Well over a trillion dollars has been invested in America's grid.

Power of the purse

Utilities wouldn't be allowed to exploit or extort customers to extract excess profit. The railroads and other trusts had done that in recent memory. So this was a vital doctrine of utility regulation.

The regulatory process ensured the abuses of the trusts wouldn't happen. A utility's regulators, not the utility itself, would set grid electricity prices, henceforth referred to as "rates." Regulators have, ever since, held the power of grid electricity's purse.

The utility does propose a budget of its "revenue requirements," just as governors do for state government. But regulators set the budget, just as the legislature does for a state. Regulators and legislatures determine budgets, not utilities and governors, whose proposed budgets are sometimes considered dead on arrival.

Based upon the regulator-set budget, the utility does propose rates, just as a governor proposes state taxes. But regulators set rates, just as the legislature sets taxes for a state.

The utility carries out the regulator-set budget, funded through regulator-set rates. This is just as the governor does for the legislature-set budget, funded through legislature-set taxes. So, utilities and governors are administrators carrying out the will of regulators and legislatures.

48 Richard F. Hirsh, *Technology and Transformation in the American Electric Utility Industry*, Cambridge University Press, 1989. "Civic reformers, trust-busting politicians, and muckraking journalists supported the public power movement during an era when "big business," as typified by John D. Rockefeller's Standard Oil Company, acquired negative connotations... [Chicago Edison Company president Samuel] Insull knew that he would need to garner popular acceptance in order to win monopoly status for his utility... Insull advocated as early as 1898 that private companies make a bargain with state governments; in return for exclusive production and distribution rights within a region, the private utility would accept regulatory oversight by a governmental body."

Utilities and governors have an important role to play. I'm not belittling them. But the powers of the administrator role also shouldn't be exaggerated. Neither utilities nor governors are kings in their realms. Neither have budgeting, pricing, or taxing authority. Those powers are held elsewhere.

Essential

It takes a year or so for regulators to set a utility's budget. It's a mind-numbing process. Typically, there are hundreds of legal filings. There are dozens of technical experts with their testimonies and tables. There are court-like hearings, a steady stream of them. Finally, there are ponderous deliberations by regulators. You wouldn't believe the extreme detail and precision that rivets these proceedings. Can the detail and precision get out of hand sometimes?

Take a look. A single rate-case issue, over amounts in the thousands of dollars, can provoke the passions of the many participating lawyers. Thousands of dollars sure sounds like a lot until you realize that, if approved by regulators, it'll move the utility's budget a millionth up or a millionth down. It'll move a household's utility bill a few ten-thousandths up or a few ten-thousandths down.

The feistiness of the lawyers is not deterred by how small an issue can be for utility customers. A lawyer for an organization in the business of fighting utility proposals is just as outraged and just as indignant. His pleadings seethe and punch just as hard.

The budget, when finalized by regulators, is no-frills, bare bones. It covers only those costs demonstrated by the utility and deemed by its regulators as essential.

A cost claimed by the utility as essential isn't unless there's a ton of evidence and unless the evidence is still standing after an onslaught by attacking lawyers. The metaphor of the piñata at a child's birthday party comes to mind. In a major rate case, the piñata is the utility's evidence for costs the utility believes should be in its budget. Or perhaps the piñata is the utility itself.

And what's considered to be essential? This is the ultimate question in utility regulation. Most of the many millions of pages of lawyer pleadings and expert testimonies in US electric utility regulation, each and every year, are arguments over the answer to this question.

The answer is, at first glance, straightforward. Getting to the answer in the sixty or so major rate cases annually shouldn't require millions and millions of arcane pages.

A cost, one that regulators should approve for inclusion in a utility's budget, should be essential or not. Or perhaps a cost should be essential if it's set at x dollars but excessive if it's set at 2x dollars. That doesn't sound that hard. What's all the arguing about?

In all those major rate cases, lawyers, consultants, engineers, accountants, etc. from all of the sides, the utility side and the many sides arrayed against the utility, are being paid to argue. So they argue.

A utility engineer is being paid to argue that a project and thereby its costs are essential. A regulator staffer is being paid to argue that a depreciation method is too generous to the utility, and thereby the associated costs are inessential. A consultant is being paid by an association of local businesses to argue that the costs of a home energy efficiency program are inessential for business customers; hence rates should be raised more for homes than for businesses. A lawyer for the state's consumer advocate doesn't like that idea and is also being paid to argue that some of the utility's profit margin is inessential.

It helps that what's an essential cost for inclusion in the utility's budget and what isn't is highly subjective. You can argue endlessly about whether a cost is essential or not. And that's exactly what everybody does in major rate cases. Essential, it turns out, is in the eye of the beholder.

The ultimate judge, the utility's body of regulators, has the final say. Some costs it will deem essential for inclusion in the utility's budget. Other costs will be deemed inessential.

Customers' rates are based on the essential costs only, those that make it into the utility's budget. So regulators are extra attentive to the bottom line. What will the budget total up to? Most importantly, what will that mean for rates?

Notwithstanding all of the arguing about whether this or that cost is essential or inessential, what's going to be the end result? How much will rates increase? How angry will those in the business of fighting utility proposals act? How will political leaders and the media react?

Not a penny more

A regulator-set budget should enable the utility to deliver a basic level of electric service to all customer premises within the franchise territory. That's all. The utility is equipped to accomplish nothing more than that and nothing less.

Sure, a utility might like its regulators to approve non-essential costs for inclusion in the budget. Ye Olde Electric Company might like its regulators to approve a higher margin and might like to scrimp on service, all to pump up profits. But as much as utilities might like these things, regulators are deputized by state legislatures to say nope to them all.

No investment in new assets is approved unless first shown in the court of public opinion to be essential and not excessive. Nor is investment in existing assets approved unless first shown in that same court to be essential and not excessive. No gold plating permitted. Unless and until regulators adjudge investments to be "used and useful," customers are not made to pick up the tab when paying utility bills.

Customers' rates are to be precisely set by regulators to compensate utilities for essential operating costs and investments but not a penny more. Not only that, every utility dollar that is spent is to be meticulously recorded by utilities and audited periodically by regulators.

Hell to pay

On occasion, regulators have found a utility spending money on stuff that is not used and useful for a basic level of electric service. There was hell to pay. So utilities are naturally gun-shy about any such expenditure. They're even gun-shy about borderline cases in which their expenditures aren't absolutely used and aren't absolutely useful.

Critics gleefully deride utilities for being risk-averse, for being wary to take chances and spend in the pursuit of innovation. Any wonder why? Some of those same critics have just as gleefully harassed utilities for being risk-prone, the opposite tendency, for taking chances and spending that didn't produce indisputable benefits.

Some have said that utilities are continually manipulating and deceiving regulators and have from time immemorial. They claim utilities routinely hide unneeded expenditures from regulators, expenditures whose only purpose is to prop up customers' rates and utilities' profits. The inference is that utility regulators are inept or stooges of utilities or both.

Come on. Any cross-section of public servants is fallible and has a few bad apples among them. This is true of any profession from carpenters to corporate bigwigs to the clergy. That being said, utility regulators are armed with tough weapons for oversight and enforcement. Almost all regulators are highly trained and motivated and independent of any untoward influences whether from utilities or from utility opponents.

Utility regulators have an unfettered license to audit utilities' books. A utility employee might foolishly seek to hide unneeded expenditures from regulators. But there is so much transparency that the foolish behavior will almost surely be discovered. And the consequences will be, for the employee and the utility, catastrophic.

The deterrence in utility regulation is fearsome. This is well understood by every regulated utility management. When a utility executive is compelled to "resign," the entire industry listens very intently. When a utility is assessed a punishing penalty for millions of dollars, the entire industry listens very intently. The same happens when a utility's franchise, its very existence as a company, is threatened.

Compelled resignations, punishing penalties, threatened disenfranchisements, all of these hammers fall with frequency. Each time the entire industry listens intently. Each time the deterrence in utility regulation is reinforced. Each time utilities are reminded of the used and useful imperative, that every expenditure covered by customers' rates must be demonstrably needed for the provision of a basic level of electric service.

A utility exec resigns in disgrace after allegations surface that he might have attempted, *might have* attempted, to influence a regulator with a job opportunity. A utility is penalized, forfeiting millions of dollars, when a project doesn't pan out. A state's governor threatens a utility's franchise and very existence with unsubstantiated allegations about the company's performance during a power outage. Each time these things happen, utilities nationwide become more and more wary.

People assume utilities are immune to financial risks because they are, supposedly, monopolies without competition. Wrong. A non-utility company can experience gradual erosion in its market value over time as competitors come to market with better offerings, better prices, or both. A utility company can experience immediate collapse in its market value anytime its regulators meet or even speak crossly about it to the media.

So do you really think non-utility companies face less risk than utility companies? In reality, non-utilities and utilities both face risk, though different kinds.

General Public Utilities, Long Island Lighting Company, and Pacific Gas and Electric are but three examples of utilities that experienced the catastrophe that riskless monopolies should have been able to elude. Their managements and stockholders suffered terribly. Every regulated utility management remembers these disasters decades later. The lessons from them guide utility behavior each day.

Many other utilities no longer exist, swallowed up by others after faltering steps accentuated by regulator disaffection. Cleveland Electric Illuminating, Illinois Power, Indianapolis Power and Light, Montana Power, and Public Service of New Hampshire were among the many utilities whose fate was their demise.

Order takers

Regulators severely restrict the business practices of utilities. It's always been so, since the dawn of utility regulation, though these limitations have become stricter in recent years. Perhaps that's because utility opponents are emboldened. Perhaps that's because utilities are defensive. Perhaps both trends are at play.

As we've seen, utilities may only provide an undifferentiated service. They may only charge an undifferentiated price. Non-utility companies, on the other hand, are free to deal, dicker, discount, differentiate, and discriminate (except against protected classes).

Utilities are, in many respects, more agencies of government than they are businesses. The greatest utility leaderships have a sixth sense about what their regulators will approve of and what they will frown upon. Other utility leaderships poorly anticipate their regulators' likes and dislikes. They tend to fare poorly and, often, they leave the scene before their time.

Utilities have devolved to disinterested order takers, passive middlemen. Utility customers take the electricity they desire without a care. Utilities must keep enough in stock twenty-four/seven and deliver it up on demand. Indeed, the relationship between utility and utility customer hardly seems at all like a business transaction between buyer and seller.

Utilities have become increasingly indifferent about their service. It's weird. Suppose customers start cutting back their purchases from utilities. Regulators and advocates of a lesser grid are prodding customers to go that way. Utilities seem to be all for that. But suppose customers start ratcheting up their purchases from utilities when new machines, appliances, and devices hit the store shelves. Utilities seem strangely silent about that.

Utilities are about the only industry in America that may not promote their product or service. The political correctness police would be too offended. Encouraging words about electricity? That would be way too edgy.

Can you come up with another industry in this pickle? Oil and coal companies may promote. Home-brew electricity companies such as solar-panel installers may promote. Other regulated companies—that sell natural gas, water, cab rides, landline telephone service, and cable television—may promote.

Somehow, electric utilities are virtually all alone. They're businesses, yet they can't market their wares. Even tobacco companies can. Even purveyors of pornography can. Even drug companies can, for legal drugs anyway.

Utilities have become neutral about their price. Customers pay what regulators decide. Utilities care enough about what the prices are. They weigh in heavily during regulatory proceedings. But so do the business-sponsored and government-sponsored lawyers and consultants who oppose whatever utilities propose, in knee-jerk fashion.

When regulators decide, the utility's input counts for no more than that of each and every opponent. Utility regulators strictly adhere to this enshrined principle of equality.

Critics have sometimes portrayed utilities as out-of-control predators preying on customers. That's crazy. If utilities were predators, they've long since been defanged. Within the business world, they're patsies, not predators. They do whatever they're told to do by regulators. Indeed, the chief executive officer of a multi-billion-dollar utility once told me, privately, that his job was boring. Does that sound like the chief exec of a predator?

Soviet toothpaste

A utility's monopoly market is presumed to be homogeneous. Its single service aims to be one size fits all.

That customers varied in their wants and needs from the grid, even back then when utility regulation was formulated, was largely ignored. For the first commandment of utility regulation was: thou shalt not discriminate. Utility service and pricing was to be doctrinally "non-discriminatory."

No one customer or group of them could be treated any differently. Like the constitutional principle of equality before the law, a utility kingpin wouldn't be allowed to advantage some customers over others.

Some households and businesses were urban, some rural, and some in between. Suburban hadn't become a term as of yet. Some were located near grid facilities, so utility costs to serve them were less, and some were not. Some were highly dependent on electricity, and some were not. Some were wealthy, so utility bills were insignificant to them, but most were not.

One size fits all. There is one uniform quality of grid electricity service. No one receives a better and more consistent quality of alternating current than anyone else, or isn't supposed to. No one gets fewer or briefer power outages than anyone else, or isn't supposed to. No one gets their customer service calls responded to faster than anyone else, or isn't supposed to.

There is one uniform price of grid electricity service. No one pays less for a given quantity of power than anyone else, within a broad class of customers, or isn't supposed to. The idea is if two households use identical quantities of power in a month, the quality of service will be identical, as will their bills.

It's as if American electricity was like Soviet toothpaste. With a single product to satisfy all of the comrades, whether Boris or Ivan liked it or not.

End of history

How times have changed. As we near the end of utility regulation's first century, states' goals seem so different.

States no longer seek to accelerate utilities' capital investment in the grid. Some political and regulatory leaders even hope investment will stagnate or rather will decline as we narrow the grid's scope in American life.

The presumption is that with the massive investment of the twentieth century behind us, the grid has been largely completed and bequeathed to us. Only marginal investment will be necessary or desirable from here on in.

The grid's growth spurts are done with, some believe. It's great that hydroelectric power plants were built in the 1930s and 1940s, that coal plants were built in the 1950s, 1960s, and 1970s, that nuclear plants were built in the 1970s and 1980s, and that natural gas plants were built in the 1990s.

It's great that the high-voltage transmission network was built in 1950s, 1960s, and 1970s for the most part. And that the low-voltage distribution network was built then and later in the twentieth century in our fast-growing South and West.

But that was then, and this is now. How times have changed.

In the last twenty years, precious few power plants of scale have been built. This generation has added almost no hydroelectric, coal, and nuclear plants of any significance. We've added some natural gas plants for the country's base-load (around-the-clock) needs but few relative to the number of aging plants built by prior generations.

Yes, a surge of wind turbines and solar panels have been put up in the last twenty years. Yet people are always shocked to hear it hasn't amounted to all that much. These renewables are really cool but really underpowered in terms of pumping out electricity.

Wind power provides just one-fiftieth of the country's grid electricity. One-fiftieth is not a lot. Solar power provides just one-thousandth. One-thousandth is basically zero. Forty-nine-fiftieths of the country's power comes from hydro, coal, nuclear, natural gas, etc., notwithstanding the hype for renewables.

In the last twenty years, precious few high-voltage transmission lines of scale and miles of low-voltage distribution have been built. We've added some high-voltage transmission lines but few relative to the number of aging lines built by prior generations. Many of these were erected to wire up wind farms.

Our recent track record of neglecting the grid speaks volumes for some. The era of neglect actually encourages them. It's the evidence they've been looking for.

Their stance is that the twentieth-century grid will be just fine for a while longer. What our fathers and grandfathers put up was, evidently, sturdy enough. We thus have the luxury of focusing our attentions elsewhere.

The expectation, or the hope, is that America can keep riding the wave ad infinitum of twentieth-century investments in power plants and wires. After all, our lights are still on, mostly.

Trust

So much is different than in those early days when states first developed utility regulation. For one, today's utilities show no inclination to menace customers as the trusts of the late nineteenth century did. Even if contemporary utilities were so inclined, it wouldn't much matter. They haven't the opportunity to become monster monopolists.

That was a realistic fear eighty years ago that utilities could become abusive like J. Pierpont Morgan's Northern Securities Company. But it cannot be a fear today.

Any such behavior by a utility now would be laughable. Nearly every aspect of a utility is so thoroughly overseen, and then second-guessed, and then third-guessed. Major rates cases are invasive colonoscopies of utilities. Nothing is hidden from the examiners.

A state's utility regulators can have as many as several hundred lawyers, accountants, economists, and engineers on staff, equipped with computers and spreadsheets. The first state regulators of eighty years ago had a handful of assistants, equipped with adding machines and ledgers.

The pioneers of regulation could not have imagined the vast resources to police utilities that their successors would inherit. They could not have envisioned how unlikely monopolistic behavior of utilities would become in the twenty-first century.

A utility has no potential, were it so inclined, to extort customers, to exploit its position as a monopoly. A utility has no potential to impose excessive costs, or to downgrade the quality of its service, to extract extra profit. The misdeeds would be reported on social media within minutes.

By this measure, utility regulation has been an enormous success, one that has gone virtually unnoticed. We should remember and celebrate this accomplishment. The original objective that states had in mind when they coupled regulating utilities with franchising utilities has been met. The threat that franchising utilities would beget monopoly abuse has been eliminated.

Utilities, like regulators, aren't perfect. But they also aren't bullying capitalists that do as they please in the pursuit of maximum profits.

Medieval tournament

Utility rate cases are all about customers' rates, at the expense of any of their other concerns. Indeed, we call these proceedings "rate cases." They're literally docketed by regulatory commissions as petitions by utilities to increase rates.

We don't call these proceedings "service cases." We don't call them "value cases" or "quality cases." Regulatory commissions don't docket them as petitions by utilities to increase service, or to increase value and or to increase quality. These proceedings aren't docketed as petitions to maintain service, to maintain value, or to maintain quality. They're all about rates.

When a state commission holds court for a rate case, there are quite a number of organizations that regularly participate. The state attorney general's organization may be there, along with the state consumer advocate's organization.

The staff of the state commission will also be there, playing a unique double role. On the one hand, the staff helps the commissioners decide the case. On the other hand, the staff takes an active side in the case against the utility joining the state attorney general and state consumer advocate in that mission.

The commission staff's conflict of interest is resolved through erecting a legal wall. There's no talking for the duration of the case between the staffers helping the commissioners decide the case and their co-workers taking an active side against the utility.

Government can have still other organizations aligning in the battle against the utility. It is typical for the federal government to fight utilities as if it were a customer. That can happen if it has military bases, government buildings, etc. pay utility bills too.

Every one or nearly every one of these groups has the same principal goal. Having so much in common, they typically support each other and ally to gang-tackle assertions by the utility.

The interveners, as they are called, have much in common. They're there to fight and suppress any customer rate increases. None have as its principal goal to fight and suppress any potential degradation in customer service, in value, or in quality.

Rate cases so single-mindedly pursue the lowest customers' rates. For the nerds out there, you could mathematically describe a rate case as an optimization. A case's goal, the objective function, is to minimize customers' rates. Minimizing rates is subject to just a few conditions (constraints, in math talk). Aging equipment shouldn't obviously threaten safety or frequent and lengthy outages. Rates shouldn't be minimized that far.

Why is rate minimization the case's objective? Where in the Ten Commandments is it written "Thou shalt not raise rates?" I haven't seen it there nor have I in the more recent public service laws. But the regulatory process presumes minimizing customers' rates is its purpose in life, its reason for existence, its raison d'être.

What if a case centered on a utility's service, value, or quality as much as on its rates? What if regulators docketed the case as a petition by the utility to increase or maintain service, value, or quality? Would we still call it a utility rate case? Or would we call it a utility service case, a value case, or a quality case?

Noisy, acrimonious quarrel

It's an obscure regulatory term. We usually write and pronounce it as an acronym, ROE. We really should sound out the acronym. It would then come out sounding like ROW, as in a noisy, acrimonious quarrel. How apt, considering the regulatory warfare that breaks out whenever the topic of a utility's ROE is bridged.

ROE stands for return on equity. It's the profit margin that regulators allow utilities when setting utility budgets and rates.

Utilities care about their ROE, a lot, for good reason. If regulators decide that ROE should be relatively low, the company's earnings in coming years will be relatively low. If, on the other hand, regulators decide that ROE should be relatively high, the company's earnings will be relatively high.

So utilities have seemingly tried everything to bolster their arguments for a higher ROE. Little has worked and nothing consistently.

And why should it? Regulators can make an easy call. They can simply choose a lower ROE. The utility will squawk. So what? So the utility's earnings won't be so great. How bad can that be?

Rate cases approach ROE like the chant of a limbo dance, "How low can you go?" It's unsurprising that regulators have responded with athletic bends and dips.

The fact of the matter is that regulators generally don't care much about how low ROE goes. There appears to be no downside.

Many of today's regulators don't care much about utilities' investment in the grid. They buy into the worldview that the grid was built out sufficiently in the twentieth century and that it needs little tending to in the twenty-first. So it shouldn't matter much that a higher ROE makes it easier for utilities to invest, and a lower ROE makes it harder.

Indeed, for those who want to squelch utility investment in the grid, driving down ROE is an effective tactic. Squeeze the profit out of utility investment and we'll see little of it. That's a desirable end state for some.

Utilities have seemingly tried everything to bolster their arguments for a higher ROE, but they've neglected the only argument that matters to utility customers. What will a higher ROE do for me? Why should I pay a few more pennies daily so you, my utility, can make more money? What's in it for me?

The answer is actually rather straightforward, bereft of the complexity of most ROE arguments. A higher ROE keeps Wall Street onboard for investments. A lower ROE throws Wall Street overboard.

What gets lost in the technical minutia of discounted cash flows and comparable companies is that Wall Street will abandon a utility when regulators trim ROE, in a New York minute. The Street has myriad investment opportunities to go to. It has zero sentiment or loyalty to a utility with a pared ROE. If regulators put a utility on the path of lower earnings, lower than other utilities and the many non-utility investment opportunities out there, it's hasta la vista, baby.

So it boils down to whether utility customers and regulators want the utility to have the Street onboard or overboard for investments in the grid. If they don't really care, then man overboard and so what? If they do care, if they do want the utility to invest to strengthen the grid, then throw down the life preservers.

It's ironic that ROE provokes such violent clashes in rate cases. For whether the regulators decide the issue one way or the other normally makes the difference of only a few pennies daily for most households. Amazing, isn't it? Whether the ROE is fifty basis points one way or the other, half a percentage point, means little to utility customers in pennies per day. Is that worth all of the intensity of today's ROE fistfights?

Allocation

Suppose a utility's regulators, after a yearlong rate case, decided the utility budget should be $3 billion. Yes, it's a big number. But the budget has a big job to do. So it makes sense the budget isn't small.

That's not all that the regulators decided in the exhaustive exhausting case. They figured, in this example, that businesses should pay two-thirds of the budget, $2 billion. Households should pay the remaining one-third, $1 billion.

How did the regulators come to that split? Could they have come out differently? Should they have?

The arcane arithmetic behind it all is referred to as "cost allocation." The concept goes like this. Take Markdown Electric Company (MEC, but everybody calls it Meek). After hearing ad nauseam from the accountants of Meek and of the organizations aligned against it in the rate case, regulators decide that two-thirds of the utility's essential costs are "caused" by businesses taking grid electricity. It's as if, if Meek didn't have to serve any of the households in the area, this two-thirds of the costs would still be necessary to serve all of the businesses.

In so doing, regulators decide that one-third of the utility's essential costs are caused by households taking grid electricity. It's as if, if Meek didn't have to serve any of the businesses in the area, this one-third of the costs would still be necessary to serve all of the households.

Because the overall total of the utility's essential costs are $3 billion, the split goes two-thirds to businesses and one-third to households. Businesses, the twenty thousand industrial and non-industrial companies and organizations, pick up twice as much as do the million households. Seem out of proportion to you?

To the regulators, it seemed fair. A business takes much more grid electricity, on average, than a household takes. So the average business should, by rights, pay bigger bills.

A large household can take more from the grid than a small business takes. But that's an exception to the rule. Generally, a business takes much more.

A business also makes much greater expenditures to keep it going, on average, than a household makes. So the average business should, you would think, more easily pay bigger bills.

A large household can expend more than a small business expends. But, again, that's an exception to the rule. Generally, a business expends much more.

This same split would typically apply to rate increases. Suppose the regulators decide they need to up the utility's budget by 10 percent, $300 million. Then businesses will be picking up two-thirds or $200 million of the new tab. Households will be picking up one-third or $100 million of the new tab.

Rates by design

With cost allocation behind us, it's time to move on to the companion arithmetic of "rate design." In cost allocation, the regulators had split up the utility budget between businesses on the one hand and households on the other. (There's some complexity we're glossing over, like that industrial and non-industrial businesses are separated out at some point. But bear with me.) In rate design, the regulators will slice and dice up the budget further among businesses and among households.

Having done that, there'll be some math and machinations to come to grid electricity prices, what we commonly know as rates. When finally reaching the end of the whole tortuous process, we'll have rates for every kind of business and household.

You cannot find it surprising then that these rates are as complicated as any prices consumers encounter in their daily lives. Regulators started out eighty years ago setting simple rates. But over the years they've added more and more complexity.

A household's monthly utility bill now rivals its monthly cellphone bill in complexity. Both of them are jam-packed with obscure line items. Which line item is more disconnected with how you live your life, kilowatt-hours usage over one thousand or data usage over two gigabytes? It's hard to say. Grid electricity rates for businesses are even more complicated, so much so that businesses of any size have someone specialize in utility bill sleuthing.

Because both businesses and households generally take all of the grid electricity they need and want with relatively little regard for its relatively low prices, prices drive utility bills. If regulators increase prices (or rates) 5 percent higher across the board, everyone's utility bills will be about 5 percent higher than what they would have been without the increase. If regulators set prices 10 percent higher, everyone's utility bills will be about 10 percent higher than what they would have been without the increase.

Because prices have so little effect on how much electricity we take, rates, bills, and utility budgets move in unison. Regulators raise rates by 10 percent confident that they're raising enough money for the utility for it to pay for a 10-percent bigger budget.

So rates and utility budgets are closely coupled. That's why a lawyer whose job it is to oppose rate increases is so feisty about this or that piece of a utility's budget. The lawyer knows that any time regulators nudge up the budget, rates will nudge up too.

Recall that Meek serves twenty thousand businesses. Regulators have set prices so the monthly utility bills these businesses receive over the course of a year total $2 billion. Okay, it won't come out exactly. Some businesses will close up shop, others will open up, some will cut back their electricity usage, and others will ramp it up. But that's the goal: have the businesses kick in $2 billion.

That comes out to an average annual bill per business of $100,000. This means businesses' monthly bills will average $8,333. And it means businesses' grid electricity costs will average $278 per day.

That can be a lot for the corner snack shop but not a lot for the big steel mill outside of town that takes much more electricity from the grid. The shop may pay two dollars daily for grid electricity. The mill may pay $25,000 daily. It's quite a difference. But when all of the twenty thousand businesses served by Meek are counted in, they average $278 per day.

Recall that Meek serves a million households. Regulators have set prices so the monthly utility bills these households receive over the course of a year total $1 billion. Again, it won't come out exactly. Some households will move away, others will move in, some will cut back their electricity usage, and others will ramp it up. But that's the goal: have the households kick in $1 billion.

That comes out to an average annual bill per household of $1,000. This means households' monthly bills will average $83.33, in the summertime more and the rest of the year less. And it means households' grid electricity costs will average $2.78 per day, in the summer more and the rest less.

That can be a lot for an elderly woman on a fixed-income who lives in a small apartment but not a lot for an affluent family of six who lives in a large house. The elderly woman may pay one dollar daily for grid electricity. The affluent family that takes much more electricity from the grid may pay six dollars daily. It's quite a difference. But when all of the million households served by Meek are counted in, they average $2.78 daily.

Because the twenty thousand businesses average grid electricity costs of $278 daily, if regulators up the utility's budget by 10 percent, and with it prices, then this average will rise by $27.80 per day. For the shop paying two dollars daily, the rise will be twenty cents per day. For the mill paying $25,000 daily, the rise will be $2,500 per day. The variety in effects from the price increase will be as broad as the variety in businesses' electricity usage and their bills pre-increase.

Because the million households average grid electricity costs of $2.78 daily, if regulators up the utility's budget by 10 percent, and with it prices, then this average will rise by 27.8 cents per day. For the elderly woman paying one dollar daily, the rise will be ten cents daily, a dime. For the affluent family paying six dollars daily, the rise will be sixty cents daily. As with businesses, the variety in effects from the price increase will be as broad as the variety in households' electricity usage and their bills pre-increase.

Tweaking

All this assumes that the regulators don't fiddle with the cost allocation or the rate design. Though they can. It's within their power, within some limits, to tweak the split between what businesses and households pay, the split between what small and large businesses pay, and the split between what small and large households pay.

The regulators of Meek could decide to have households pick up a little bit more of a rate increase than one-third, to keep businesses' grid electricity costs from rising by much. This might be a good regulatory policy to attract greater business formation in the area. With greater business formation would come more jobs and then greater household formation. Such economic development is exactly what public service laws envisioned, when written eighty years ago or so, as an overriding objective. Indeed, public service laws often specify the objective of economic development explicitly.

The regulators of Meek could decide to have large households pick up a little bit more of a rate increase to keep small households' grid electricity costs from rising by much. That elderly woman could be assessed a rise of less than ten cents per day while the affluent family is assessed a rise of greater than sixty cents. This might be a good regulatory policy to lessen the impact on those households least able to pay more. With a lessened impact on these households, regulators may feel freer to set adequate utility budgets fully up to the task of seeing to it that power outages for all are rare and brief.

Graph 16:

High-income households spend more than low-income in all categories but difference not that much in electricity.

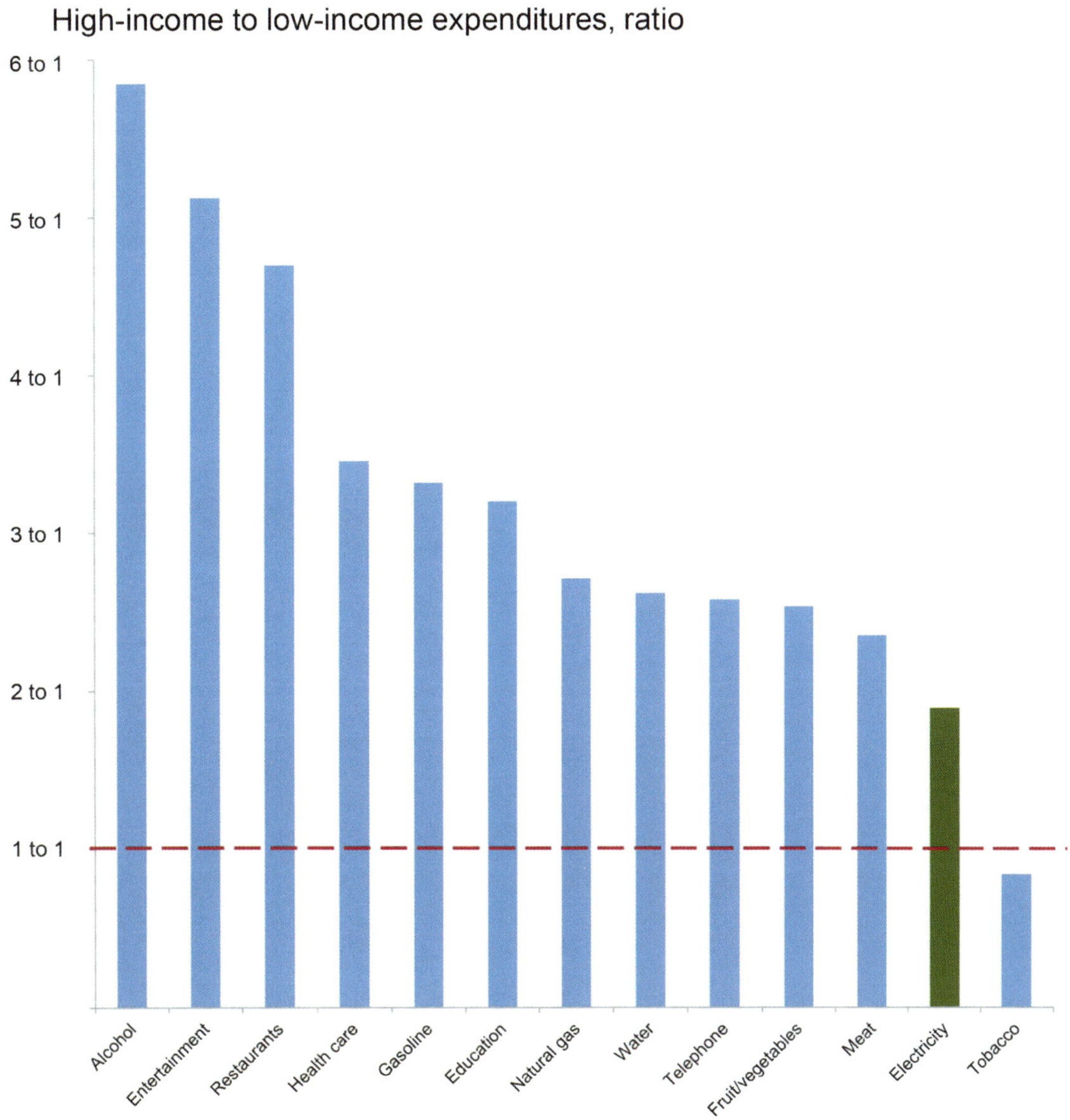

Graph 17:

Outside of California, electricity prices paid by high-income and low-income households are nearly equal.

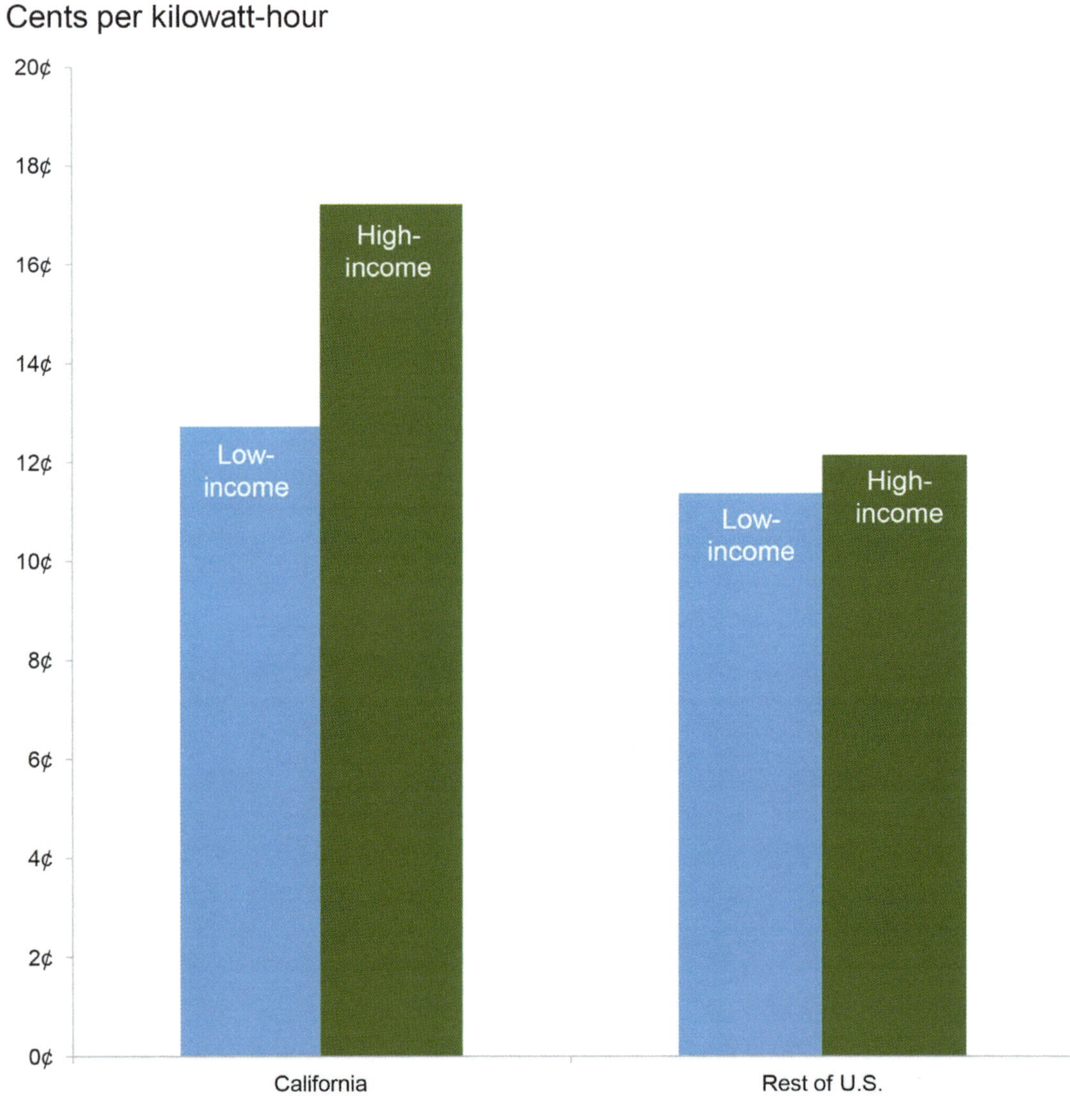

You're testifying on what?

There hadn't been a rate case this big for a couple of years. The utility petitioned its regulators to increase its budget by hundreds of millions. If regulators went along, rates would rise. The opposition geared up and shifted into overdrive. The lawyers whose job it is to oppose rate increases were outraged, the media reported.

The opponent pros lobbed the tried and true barbs and bombs. The rate increase will spread ruin like when the Huns swept through Europe. Local companies will go belly-up. Families will be forced to choose between food and fuel. The elderly will not have enough money for medicine.

The utility could have done what utilities do these days. Drop into a defensive crouch. Apologize for asking for an increase. Show remorse for any harm it might have. No matter how hard opponents hit, no matter how much they trash the utility's truthfulness and good intentions, stay calm and cool. Remain professional. Stick to contesting the technicalities of the increase.

The utility didn't stick with the same old script. It dared to air a positive campaign that spoke straight to its customers. The focus was on the utility's extraordinary value to customers, both before the increase and afterwards. This approach was unprecedented. No utility had tried that.

Then a funny thing happened. It took place even as the opponent pros kept up their onslaught of attacks and accusations. It took place even as they positioned the increase as no more than selfishness on the part of the utility to push up profits.

Households and businesses responded. They generally accepted that the utility's value had been exceptional. And they were skeptical about the opposition's evident exaggerations. The utility found this in surveys and one-on-ones with customers.

Nonetheless, the more aggressive opponents were stuck in attack mode. Their testimonies in the case were seasoned with some reckless statements. They had, it turned out, overreached.

The utility departed again from the tired script. It dared to submit rebuttal testimony that spoke simultaneously to customers and regulators. The focus was, again, on the utility's extraordinary value, before the increase and afterwards. The rebuttal demonstrated quantitatively, with actual customer bill data, that the increase's impact on households would be only a few pennies daily. The impact on businesses would be small as well.

The whole dynamic of the rate case changed. What could the remaining opponents say? If the utility's value was exceptional and its cost remarkably affordable, what could they say?

The remaining opponents harped about this and that technicality. But their arguments seemed trite at that point. The big picture, from the perspective of utility customers, was that the utility's value proposition was pretty good, even if the increase was instituted.

And the increase was. The utility collaborated and compromised somewhat with the more circumspect opponents. The regulators were clearly pleased. They voted unanimously to approve the compromise.

The utility gets to invest hundreds of millions to enhance the value that customers get from the grid. Households and businesses get that value for an increase that's barely noticeable. The regulators get the satisfaction that they steered an unwieldy ship toward a promising destination. As for the most aggressive opponents, perhaps they gained, too, seeing a different side of what utility regulation should be about.

Reddy's Recharges

1. At the dawn of utility regulation, states awarded utilities monopoly franchises to accelerate electrification and, at the same time, established permanent commissions to ensure the new monopolies wouldn't run amok.

2. The combination of franchising and regulating has been an enormous success, prompting utilities to invest well over $1 trillion in grid electricity while preventing utilities from abusing their unique position.

3. But the regulatory process has, in the last twenty years, promoted more criticism than collaboration, cowing utilities into inaction, immobilizing them, and discouraging innovation to increase (even maintain) the value of electricity the public wants and needs.

"I took him for what he was worth, but it was all in utilities."

Chapter 9: Enter Lengthy Storm Outages

Storm outage impacts; restoration distribution skewed too; value losses; utility, regulator options; bill impacts

Nineteenth nervous breakdown

A power line is art. Strung between mighty poles, the wire is slack, slinky, and still. But our eyes are drawn back to the poles, the wooden pickets along roads and streets.

Then the weather turns. We look up. The wire sways. It's a graceful dance as the music of a storm starts up.

A tree branch was spared in the last trim. It falls. The limb is soaked heavy. On the way to the ground, the branch...

The jostling is just enough. The wire is but an infinitesimal link in an infinite galaxy of electrical equipment. But if the wrong link is jostled...

The wire's current, now flummoxed, looks wildly for other ways home. Other startled currents are shoved aside. Now there are violent stampedes all about. What was once a jostle in northern Ohio has become crazed riots throughout the Midwest, northeast United States, and eastern Canada.

I was in another world at the time, August 14, 2003. In Italy, trying to catch my breath from an impossible task. Build two super-circuits to supply New York City from the state's northland. A *New York Times* editorial called the project a modern-day Erie Canal. Many felt we could build it. Cynics knew we wouldn't be able to.

My seventy-eight-year-old dad called from New York City. What the heck is going on? The great Gotham had been brought to its knees. Subways froze in their horizontal caverns, elevators in their vertical towers. Dad hitched a ride at the mouth of the Lincoln Tunnel back to New Jersey. From and to darkness, both sides of the Hudson River were eerily the same.

Not just New York City but eight states and Ontario, too, blacked out. Within minutes from when the sodden branch fell in Ohio, over 250 power plants had slammed shut their gates lest the riotous currents bring their ruin on them.

Fifty-five million Americans and Canadians lost the power in their lives. The clock turned all the way back to 1880. No, it was much worse.

In 2003, there were no horses and carriages for transportation as there were in 1880. There were no whale oil and gas lamps for lighting. There were no shade porches and wide-open shutters for cooling. There were no coal and wood-fired ovens for cooking. Without electricity, and without coal and wood, warming options were few.

There were no mechanical cash registers and ledgers at shops in 2003. There were no ice chests to preserve food and not enough bills and coins to support commerce.

In 1880, you wouldn't add so many stories to buildings and so line their exteriors with glass. In 2003, without electricity, these buildings are uninhabitable. In 1880, you wouldn't place homes so far from farms and factories. In 2003, without electricity, these homes are impractical.

As the hours accumulated, we were reminded what really fuels society. It is our energy, yours and mine. But when the grid's energy shuts, our energy is soon sapped.

Then the engineers bandaged the wound. The first households and businesses returned to the twenty-first century in seven hours. The last got back in two days. The politicians followed with their elixirs. And the Northeast Blackout of 2003 passed from memory.

Humpty Dumpty

Lights flicker then fail. Fear, frustration, fury are your first emotions.

You look about. Who else is out? Are the neighbors shut down too? Is your whole community down? How far has the terror spread?

You feel helpless. The kids can't understand. What broke? Can you fix it?

It's like Humpty Dumpty's fall. The breaks are too big for one man. The king's men will put Humpty back together again, surely soon. You hope.

A utility truck passes. "Kids, the power will be back soon." Hours pass. Days pass. Patience, however much was left, passes too. You feel powerless.

Threshold of pain

A household or business is jarred by a power outage. The whole episode lasts a half hour. What's the big deal? The inconvenience is hardly devastating. It's an intermission, that's all. We wait it out.

A summertime blackout shuts down the air conditioning. But a house or workplace doesn't become unbearable in a half hour, even on a sweltering day.

A wintertime blackout shuts down the electric heat. But a house or workplace doesn't become unbearable in a half hour, even on a frigid day.

Whenever a blackout hits, we cope. When inside, we take off or put on layers of clothing to feel comfortable again. We take a breather from our daily tasks. Or we go about them undaunted.

Machines, appliances, and devices directly powered by grid electricity go inert. They turn off instantly if they were busily at work for us. They refuse to turn back on. None of our grid-based equipment turns on. In unison, the equipment walks out in a defiant wildcat strike.

But machines, appliances, and devices indirectly powered by grid electricity, with chargeable batteries, remain at their workbenches amidst the strike. They dutifully turn off and on, on command as always. Though, as a blackout's minutes grow into hours, our worry grows. Chargeable batteries need the grid electricity to recharge. Like a swimmer underwater, they must come up for a fresh breath after a time.

Suppose the episode takes an hour, twice as long. The impact on us is a bit greater. But it's over soon enough. Afterwards, all is forgiven. The same goes for a two-hour power outage, four times as long. The impact is greater; this is true. But, again, it's over soon enough.

We're reminded the grid can freeze up like our PCs. But it's nice to know the utility will resolutely reboot when a grid glitch develops.

We can generalize. The inconvenience to households and businesses increases with the duration of a power outage. An hour blackout is worse than a half hour. A two-hour blackout is worse than an hour. Is a two-hour blackout twice as bad as an hour?

For a power outage measured in minutes or even a few hours, it's crumby that there's an interruption. Shorter is better than longer, no doubt. But blackout costs hardly increase proportionally. We can bear it awhile while the utility crews mend the grid.

The pain is bearable, to a point. Then everything changes. We cross over our threshold of pain.

Make that power outage twelve hours, eighteen hours, a day, a couple of days, a week, a couple of weeks. Now, that's something quite different. There's a real threshold of pain. Crossing over it, an episode becomes a catastrophe.

What happened? What changes at this critical threshold?

The outage now devastates. Our grid-based superpowers, to control inside climate, to extend daytime light, to browse all mankind's knowledge, are now desperately missed.

Machines, appliances, and devices indirectly powered by grid electricity, with chargeable batteries, now leave their workbenches, too, joining in the wildcat strike. Chargeable batteries suffocate with no grid electricity to recharge.

They are now rendered as inanimate and useless as their co-workers, equipment directly powered by grid electricity. Like in the sci-fi classic *The Day the Earth Stood Still*, society stills.

We can again generalize. The inconvenience to households and businesses continues to increase with the duration of a power outage. A one-day blackout is worse than a half-day. A two-day blackout is worse than a one-day. Is a two-day blackout twice as bad as a one-day? For most families and organizations, it's about twice as bad or even worse.

Blackout costs now increase proportionally or possibly exponentially. Our patience has worn thin. We worry. We watch. Where is the utility? Where is the expletive utility?

The pressure on the utility, as palpable as it is during a power outage, soon passes afterwards. We pat ourselves on the back. We celebrate a job well done.

Attila

Multi-day outages land a sucker punch on only a minority of a minority of a utility's customers. These poor souls, the exceptions to the rule, are but a few percent. That's doesn't seem so bad.

In most blackouts, a majority of customers never lose their grid-based superpowers. They continue to bring to life, at will, uninterrupted, their machines, appliances, and devices. For them, a blackout is nothing but a heart-wrenching reality show about their neighbors, dramatized on their late-night news.

Let's look at the drama of a blackout played out. We zoom in on the world of a made-up utility, Invincible Powers, Inc. Invincible connects the grid to its 1.5 million household and business customers. Its motto, "You're powered by Invincible, so you can't be stopped."

Hurricane Attila, a furious force, attacks the typically tranquil area served by Invincible. As when the Huns swept through Eurasia, the hurricane punctures the peace then pillages the people. Yet, of the 1.5 million customers of Invincible, a million are not affected. Their grid-based superpowers remain intact. Lucky them.

A minority of customers does lose their electrifying superpowers. These half million are not so lucky. Their powers are temporarily gone, like Superman exposed to kryptonite. But most regain their superhuman strength within hours, in a happy comic-book ending.

Twenty-four hours after the attack, 80 percent of the half million are back to full strength. It's impressive. In just the first day, the lines feeding 400,000 customers are re-electrified. Pretty good. Only 100,000 households and businesses, one in fifteen of the Invincible customer base, remain in the storm's ruin.

Utility crews are doing the best they can. They clean up mess after mess in a logical sequence. But a minority of the minority is left powerless as the hours pile on. Forty-eight hours after the attack, 90 percent of the half million are back to full strength. Only fifty thousand, 10 percent, remain in the ruin. Those still powerless turn sour. Later they'll turn savage.

Local stations spring into action, savor the story. On-the-scene ace reporters dramatize the desperation as it's developing on darkened streets.

Seventy-two hours after the attack, 95 percent of the half million are back. Only twenty-five thousand, or 5 percent, remain off. Ninety-six hours after, 99 percent are back.

Utility crews are still hard at it, without letup. Now into day five, they've been reinforced by waves of fresh troops. Fellow linemen drove in from around the region and around the country. But it's been a string of long days of non-stop hand-to-hand combat, stringing our lines back up. The front line is exhausted.

Only five thousand, or 1 percent, remain off. Those still powerless are few in number, relatively, just one in three hundred of the Invincible customer base. But they're a real surly group by now. Their anger is magnifying their number.

The hapless five thousand were restored to the Invincible grid over the next few days. The historic Colonial of someone's grandmother, the rambling ranch of a family with four infants, the low-rise complex of cramped apartments, the row of small retailers, the struggling pizza parlor—they were among the still powerless.

The last customer of Invincible was reconnected. It was day eight after Attila attacked. The stay-at-home dad was happy to return to normalcy, though taxed by the terrible trial.

Storm skew

When Hurricane Attila attacked, a half million households and businesses served by Invincible were severed from the grid. As soon as the winds died down, Invincible personnel began counting the customers cut off, assessing the mess, and reconnecting public safety priorities.

In the first hour after Attila passed through, Invincible counted 200,000 customers cut off. By the second hour, Invincible knew the customer count from lines down was up to 500,000. This was the base, 500,000, that Invincible worked from.

By the end of the sixth hour, 150,000 of the 500,000 had been reunited with the grid. By the end of the twelfth hour, the reconnects were up to 250,000. Now half were back online.

By the end of the eighteenth hour, the reconnects were up to 350,000. By the end of the twenty-fourth hour, they were up to 400,000. Now 80 percent were back online.

To wrap up day one, 150,000 customers were restored during the first six hours, 100,000 were during the second six hours, another 100,000 were during the third six hours, and fifty thousand more were during the fourth six hours.

There was just 20 percent to go, 100,000 customers. But diminishing returns were starting to set in.

During day two, fifty thousand more customers were restored, driving the cumulative percentage restored up to 90 percent. There was just 10 percent to go, fifty thousand customers.

During day three, twenty-five thousand more customers were restored, driving the cumulative percentage restored up to 95 percent. There was just 5 percent to go, twenty-five thousand customers. It was slow going now.

During day four, twenty thousand more customers were restored, driving the cumulative percentage restored up to 99 percent. There was just 1 percent to go, five thousand customers. It was real slow going now.

Over the succeeding four days, success was a slog. Day five saw twenty-five hundred customers restored. Day six saw fifteen hundred restored. There was just one-fifth of 1 percent to go, one thousand customers. They weren't happy campers. Some were camping, without their grid electricity, in their backyards.

Day seven saw 750 restored. There was just one-twentieth of 1 percent to go, 250 truly vexed customers. And then, on day eight, the ordeal was over. Crews brought back up the last lines that Attila had torn down eight days earlier.

Do you see? Just like the statistical distributions of how much electricity customers take from the grid and how much they pay for what they take, examined in earlier chapters, the statistical distribution of storm restoration is also skewed. Like we said earlier, practically everything in electricitynomics is skewed.

In these skewed distributions, most of the action is at the low end of the scale, on the left-hand side of the graph. But a little bit of the action is way out there at very high end of the scale, on the extreme right-hand side of the graph. That right-side tail is small in numbers but disproportionally influential on the overall statistics of a distribution.

Plot the number of powerless utility customers against time. On day one, the powerless peaked at 500,000. Invincible restored 50 percent of the 500,000 by the twelfth hour and 80 percent of the 500,000 by the end of day one. If Attila knocked out your power, chances are Invincible restored it within twenty-four hours.

But for the 20 percent that Invincible didn't restore by the end of day one, their chances were chancy and situation dire. Of the 100,000 customers that Invincible didn't restore that first day, half would have an outage lasting three or more days.

The hapless fifty thousand, which would endure an outage lasting three or more days, are in the right-side tail of the distribution. Half of them endured an outage of four or more days. Those households and businesses are further out there in the distribution's tail.

The real hapless one thousand not restored until day seven or eight are way out there in the tail. They endured the Hurricane Attila blackout at its worse.

Their numbers are relatively small, one-fifth of 1 percent of the customers knocked offline originally. But the severity of their experience makes them disproportionately important, both to Invincible and the regulator of Invincible.

Graph 18:

Superstorm Sandy's skewed distribution of households and businesses without power, New Jersey

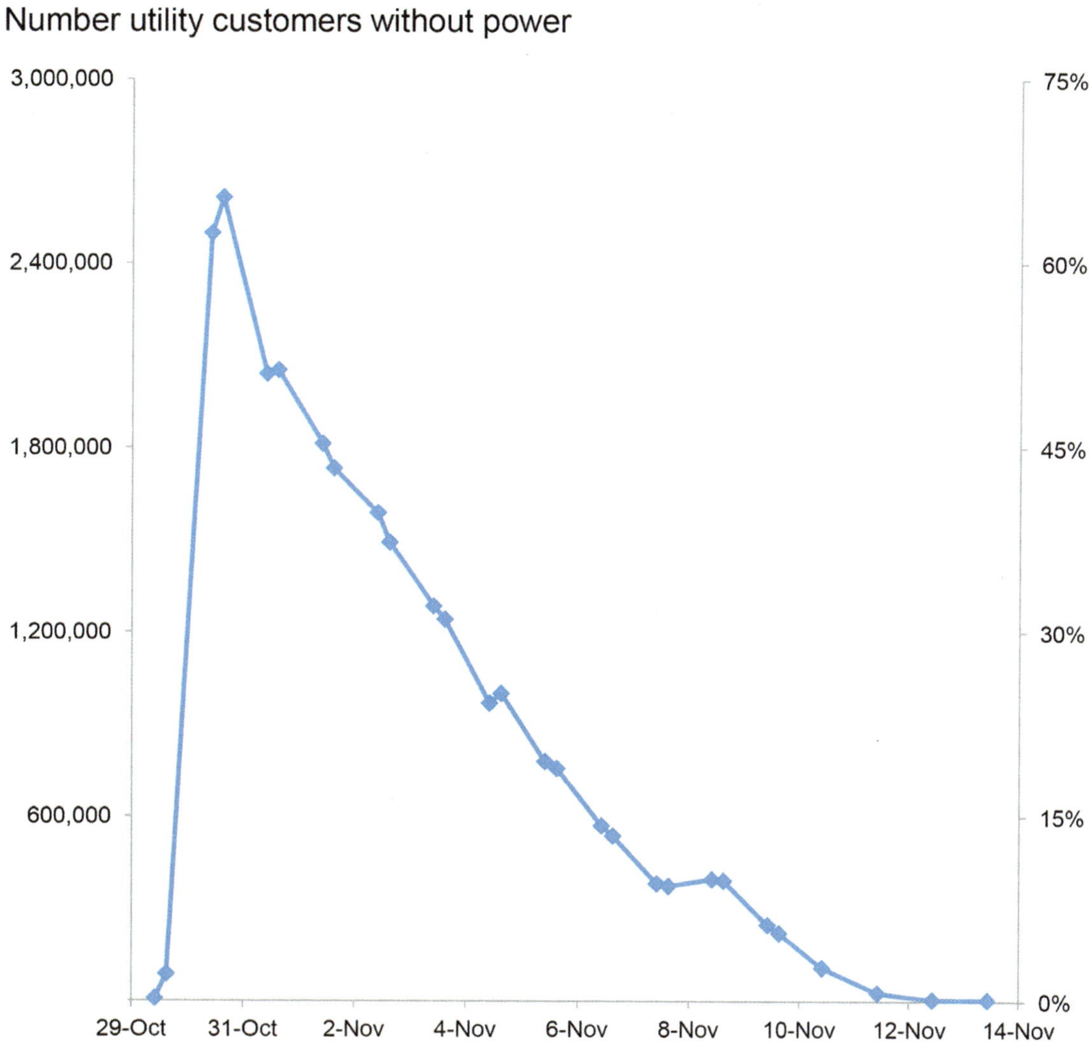

Graph 19:

Superstorm Sandy's skewed distribution of households and businesses without power, New York

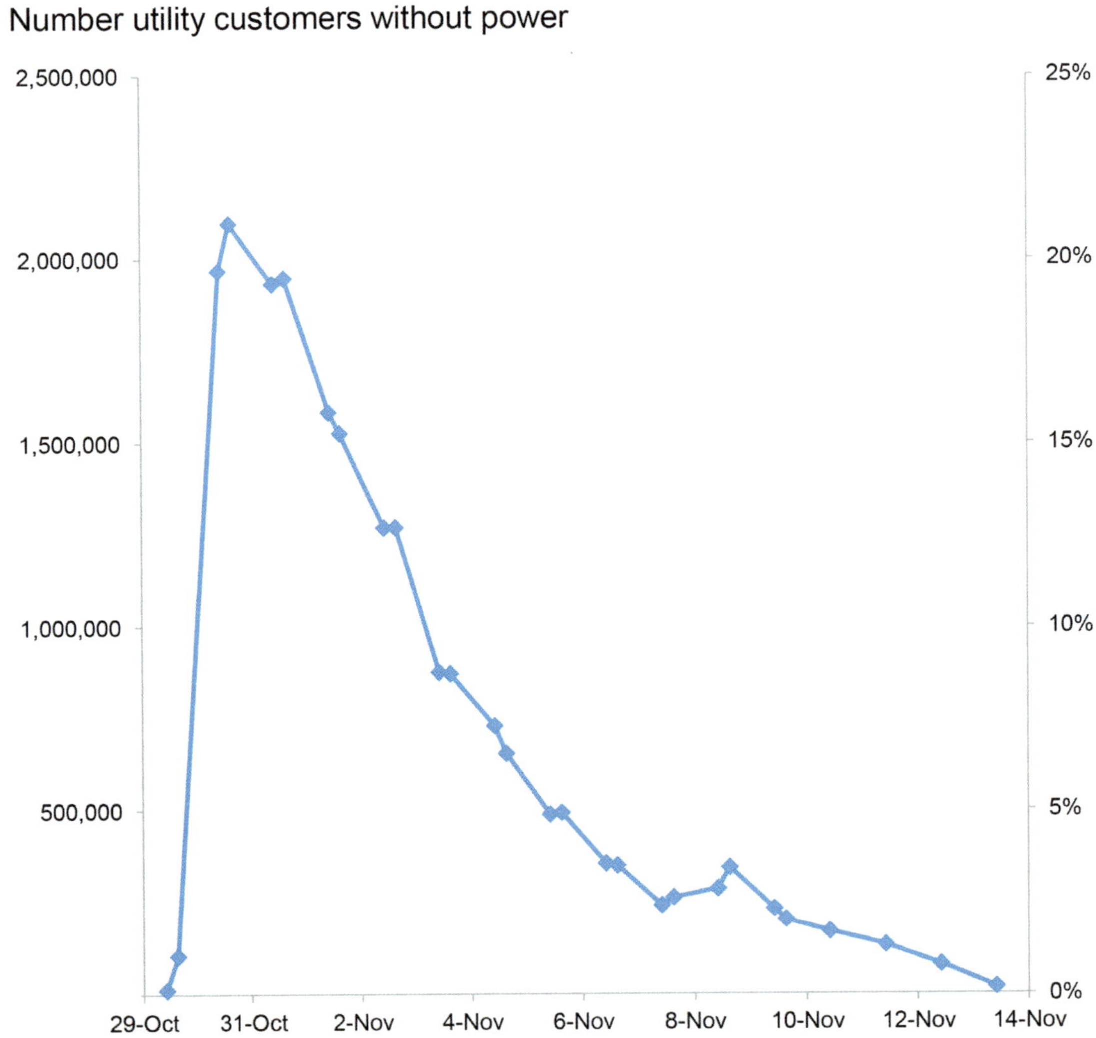

Seven hundred dollars a week

On day five after Hurricane Attila's rampage, just 1 percent of all those whose grid connection snapped in the storm remained off. Only five thousand households and businesses at that point stayed severed. They were unhappy, to be sure. But they weren't all that numerous.

Invincible was streaming power to 299 out of three hundred of its customer base. They're 99.7 percent. The still powerless were only one in three hundred. They're 0.3 percent. What's the big deal?

The big deal is that, for those five thousand, the costs of the Hurricane Attila blackout were enormous. They incurred considerable out-of-pocket expenses. They experienced severe distress and discomfort. Distress and discomfort are more difficult to quantify than expenses. Nonetheless, they are real economic costs that consumers willingly pay to avoid or mitigate.[49, 50]

Suppose the first hour of a blackout imposes a cost on a utility customer of fifty cents. It's not much, but it's not zero. Would you pay fifty cents to avoid a one-hour blackout? Sure you would.

Suppose the second hour imposes fifty-five cents. That's still not that much. The third hour imposes sixty-five cents. The fourth hour imposes eighty cents. The fifth hour imposes one hundred cents, a whole dollar.

See where we're going? The customer's outage cost function, we're supposing here, is exponential. His or her costs of a blackout increase exponentially. That just means the outage

49 Diane Cardwell, Matthew L. Wald, and Christopher Drew, "Hurricane Sandy Alters Utilities' Calculus on Upgrades," *New York Times*, Dec. 28, 2012. "And there is growing recognition that the true cost of disruptions, in terms of gasoline lines, lost workdays and business sales, and shivering homeowners, is far higher than the simple dollars and cents spent to protect the power system. A recent report from the National Academy of Sciences about the vast 2003 blackout in the Eastern United States determined that the economic cost of that disruption was about 50 times higher than the price of the actual electricity lost, and that didn't take into account deaths or other human consequences. 'We need to think now of not just restoring the grid, but how to make it more survivable,' said Philip B. Jones, president of the National Association of Regulatory Utility Commissioners, a trade association of state officials. 'I think most commissioners are coming around to that.'"

50 Juliet Eilperin, "Study: Proliferation of Home Air Conditioners Saves Lives," *Washington Post*, Dec. 23, 2012. "The installation of air conditioning in American homes is the reason why the chances of dying on an extremely hot day fell 80 percent over the past half-century, according to an analysis by a team of American researchers… A team of researchers from Tulane University, Carnegie Mellon University, the National Bureau of Economic Research and the Massachusetts Institute of Technology examined patterns in heat-related deaths between 1900 and 2004. The group found that days on which temperatures rose above 90 degrees Fahrenheit accounted for about 600 premature deaths annually between 1960 and 2004, one-sixth as many as would have occurred under pre-1960 conditions."

costs of every additional hour of the blackout increase, and these costs increase at an increasing rate.

The cumulative costs are fifty cents for a one-hour blackout, $1.05 for a two-hour blackout, $1.70 for a three-hour one, $2.50 for a four-hour one, and $3.50 for a five-hour one.

Does that sound right? You've lived through a five-hour blackout. Do you think your costs were around $3.50? Do you think they were higher? Were they lower? Everybody's different. Everybody feels a little differently about power outages. Everybody takes different strategies to mitigate the pain. But here's the acid test. Would you have paid $3.50 to avoid the experience? I'm guessing yes.

If we kept going with this exponential outage cost function, the cumulative costs for a blackout would really mount as the hours mount. The cumulative costs are $4.75 for a six-hour blackout, $20.30 for a twelve-hour blackout, $57.45 for an eighteen-hour one, and $126.95 for a twenty-four-hour one.

That twenty-fourth hour without power added $14.30 to the customer's outage costs, 11 percent of the cumulative total all by itself. The still-powerless customer is evidently frustrated now.

For the powered-down, after some number of hours and days, a blackout becomes numbing. You know it. Perhaps every day of a blackout after the first has an outage cost of around one hundred dollars. A one-day blackout is one hundred dollars. A two-day one is two hundred dollars. A weeklong one is seven hundred dollars. A two-week one is fourteen hundred dollars. This is much less than our exponential outage cost function, in which a one-day blackout is $124.95 and succeeding days are astronomical.

Does that sound right? Perhaps you've lived through really long blackouts. Do you think your costs were around one hundred dollars for a one-day one? Do you think they were higher? Were they lower? Again, everybody feels a little differently about power outages and takes different strategies to mitigate the pain. Here's that acid test. Would you have paid one hundred dollars to avoid the experience? Maybe yes. Maybe no.

Do you think your costs were around seven hundred dollars for a weeklong blackout? Do you think they were higher? Were they lower? Here's that acid test. Would you have paid seven hundred dollars to avoid the experience? Maybe yes. Maybe no.

Notably, you pay Invincible around twenty-five dollars for a week of grid electricity. If you would have paid anywhere near seven hundred dollars to avoid a week without grid electricity, and if you pay twenty-five dollars a week for the power, this must mean Invincible is giving you an unstoppable value.

If you would have paid seven hundred dollars to avoid an unpowered week, that's how much you pay Invincible for twenty-eight powered weeks. This is one way to illustrate that utility blackout costs far exceed utility bill costs. You pay utility bills each and every month. You incur blackout costs irregularly and occasionally. But when you do incur the costs of a blackout, it packs a punch.

Graph 20:

Lengthy storm outages are rare.

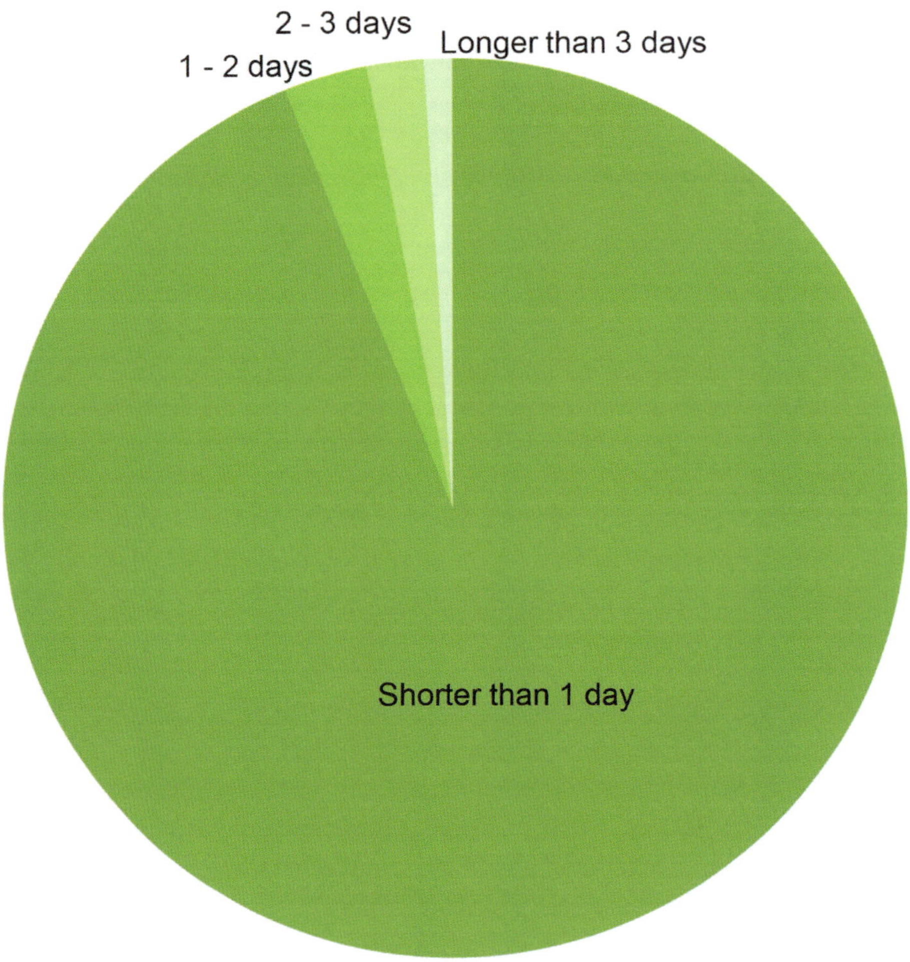

Graph 21:
Lengthy storm outages, though rare, subtract disproportionately from grid electricity's value.

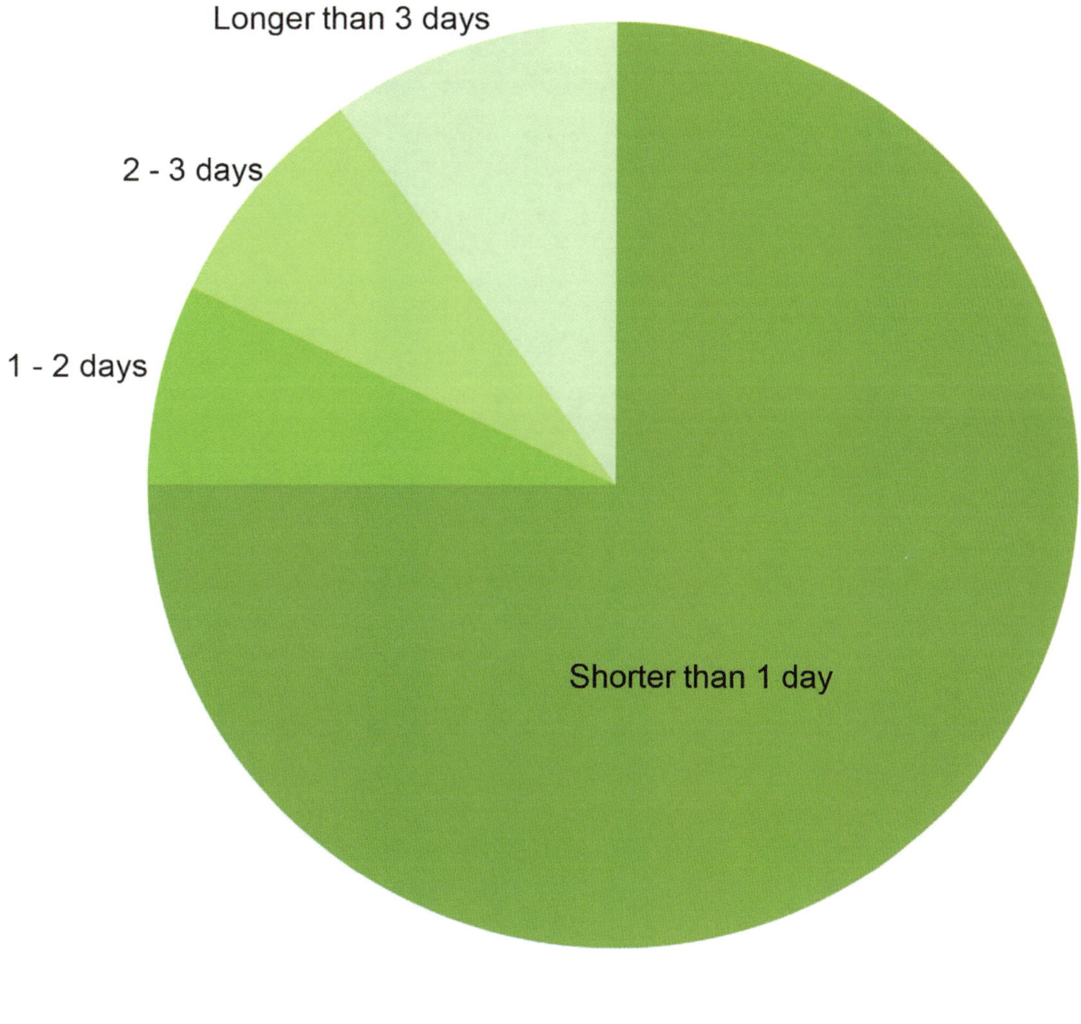

How much would you pay?

Suppose you feel that, over the course of a year, your household might incur blackout costs of around one hundred dollars. Per hour, blackouts cost you an average of two dollars. That's your perspective, but your perspective is the one that counts.

Your hourly number is just two dollars because you expect interruptions to be brief for the most part. The brief ones aren't too costly to your family.

You expect roughly fifty hours of interruptions a year. Possibly a day-long outage after a snowstorm and a few outages of a few hours apiece spread over the year. That daylong outage alone accounts for most of the one hundred dollars.

Now here's where it gets dicey. Suppose you're beginning to think your trusty utility is becoming less so. You now expect roughly one hundred hours of interruptions a year. Possibly two long outages, after a snow or summer storm, each multi-day, and a few outages of a few hours apiece spread over the year.

Per hour, blackouts now cost you an average of five dollars, per your perspective. Your hourly number is up to five dollars because you expect interruptions to be longer now. The longer ones are much costlier to your family.

Look what happened. Hours of grid interruptions increased from fifty hours to one hundred hours, a doubling. Per hour blackout costs increased from two to five dollars, more than a doubling. Per annum blackout costs increased from one hundred to five hundred dollars, five-fold more.

What if your utility, which was trusty and now isn't, returns to trustworthiness? You might believe again and expect your per annum blackout costs will return to one hundred dollars.

That's a four-hundred-dollar gain if your blackout costs come back down. Would you pay two hundred dollars for the four-hundred-dollar gain? Seems like a pretty good deal.

Your per annum bill costs, for all twelve monthly utility bills, could be around thirteen hundred dollars. If utility regulators decided to up that by two hundred dollars, a healthy 15 percent, the utility might have enough money to keep your blackouts relatively rare and brief. You'd pay two hundred dollars more a year in utility bills, fifty-five cents per day more, but avoid four hundred dollars in utility blackout grief.

So, would you do it? Would you kick in fifty-five cents per day if your utility would use the money to minimize blackouts? Most people would. That's if they could be assured their utility would indeed use the money diligently and wisely to minimize blackouts. And that's where utility regulators come in.[51]

Never again

You would expect utility regulators and utilities to try almost anything to avoid this. Why wouldn't they go all out to reduce to a rarity the incidence of multi-day outages, to truncate all outages to two or three days max, to limit this to no more than a handful of customers?

Time, talent, technology—all of these elements are brought to bear, you would expect, to fight multi-day outages. Money, manpower, management—they bring it all, you would expect. Utility regulators and utilities must have no higher objective.

Never accept leaving a poor family on public assistance unable to buy food after a storm. Never accept leaving an elderly widow in a tall apartment building unable to see her doctor. Never accept leaving a blind woman all alone unable to call for help. There can't be a higher objective in the world of utility regulation, right?

Maginot Line

Well, wrong. What you would expect is wrong.

Utility regulators and utilities, notwithstanding the dedication of repair crews, don't bring it all to the fight. Let's look at why.

For at least two decades, utilities and utility regulators have focused on an assortment of pressing priorities. Naturally, the best and brightest are attracted to work on these topics. What about the mundane task of keeping circuits circulating? That's pretty dull.

51 Patrick McGeehan, "Con Ed Seeks Rate Increase," *New York Times*, Jan. 25, 2013. About a proposal by the utility Con Ed to harden its grid against storms, at a cost to a "typical" household in Westchester County of twelve cents daily, the county executive said: "I'm sure nobody wants to pay more to Con Ed, but if the rate does go through and is used to invest in the infrastructure, bury some of the power lines and create a better contact with the customers, then at least it will be money well spent."

Deregulation and divestiture of utilities' power plants. That was hot. Developing trading markets for power plant output. That was hot.

Dampening power plant pollution. Discouraging households and businesses from taking grid electricity. Driving them to brew their own instead. Depressing utility rates. These were and are the burning issues.

Hundreds of thousands of hours in utility regulation have focused on these priorities in the last two decades. Millions of pages of expert testimony have been filed and debated. Billions of dollars have been spent on lawyers, consultants, accountants, etc. All the while, all assumed that the defense against blackouts would hold fast.

Our complacency, like France's in the 1930s, in its Maginot Line against a German attack, was sustainable for a while. Multi-day blackouts were fewer and shorter. They were less widespread. They were less disruptive and devastating.

When storms raided the frontier, the incursions were terrorizing. But the suffering from blackouts was considered inevitable. It is what it is. Anything we would try is either marginally effective, like more tree trimming, or wildly uneconomical, like undergrounding lines.

So, prepare for the worst. Ready the home emergency kit. Because, when a storm hits, it'll be lines down, lights out. That just the way it is. Better to accept that reality.

Our complacency wasn't really sustainable. Multi-day blackouts became more frequent and longer. They were more widespread. They were more disruptive and devastating.

A brutal invader has broken through impenetrable lines—impenetrable, we thought. We should have recognized the threat as it was growing. We should have reinforced the frontier in time. But, ah, too many distractions.[52]

---

52 National Research Council, "Terrorism and the Electric Power Delivery System," The National Academies Press, 2012. "Primarily because of ambiguities introduced as a result of recent restructuring of the industry and cost pressures from consumers and regulators, investment to strengthen and upgrade the grid has lagged...A terrorist attack on the power system... could deny large regions of the country access to bulk system power for weeks or even months. An event of this magnitude and duration could lead to turmoil, widespread public fear, and image of helplessness that would play directly into the hands of terrorists. If such large extended outages were to occur during times of extreme weather, *they could also result in hundreds or even thousands of deaths due to heat stress or extended exposure to extreme cold.*"

Wooden ships

America's electric grid is strung up, literally. The grid's capillaries, carrying society's vital fluid to us, hang from 150 million wooden utility poles.

Wood? Really? What's wooden anymore? Fine furniture is. Wood frames in construction are, though other materials are increasingly substituted. Major League bats are. It's mainly for the nostalgia, for the "crack of the bat." Also, metal and composite bats drive balls too hard, too far.

How is it that our society shuts down when chemically treated tree stumps fall? They're glorified toothpicks stuck in the ground. They're so fragile; storms level hundreds or thousands of them.

Wood has a glorious past. Houses were wooden. Ships and boats were wooden. Horse carriages were wooden. Weapons, tools, and implements were wooden. Bridges and mills were wooden.

But metals and synthetic materials replaced wood for almost all purposes. They're stronger. They can be flexible for resilience. They last longer and don't rot from weather nor are eaten by termites. They don't chip or splinter. Metals and synthetics are easier to fabricate exactly to the need.

So why are wooden poles our critical path? Yes, you can staple candidate posters to them. You can tack on missing cat signs. Some say they look bucolic. It reminds others of the bygone. That's nice. Wooden poles are a little cheaper than the metal and concrete poles used virtually everywhere, except the United States. And, it's said, our utility linemen like climbing the stumps.

But is that reason enough to limit the grid's survivability in storms? In military theory, survivability is a function of susceptibility, vulnerability, and recoverability. Susceptibility is the likelihood of being attacked. Vulnerability is the likelihood of being disabled when attacked. Recoverability is the likelihood of timely repair to original capability when attacked and disabled.

A wooden pole-based grid is fundamentally susceptible and vulnerable to storms.[53] Hurricanes, blizzards, tornados, etc. wreak havoc on wooden poles, tearing them up

53 Rebecca Smith, "Utilities Revisit Storm Plan," *Wall Street Journal*, Dec. 16, 2012. "[The] chief executive of Ohio-based American Electric Power Co. said utilities need to get smarter about what they install: It may not make sense to replace broken wooden poles with more wooden poles. 'We're putting up steel and concrete poles...' 'Such improvements come with a cost,' [another chief executive] said, but "it pales compared with the economic impact of power outages.""

effortlessly. As for recoverability, the best utilities can do is ship in and implant new stumps. Necessarily, that takes days.

A week?

Sitting in a Seoul hotel room, I spotted the news on my iPad. Back home in Washington, DC, and hundreds of miles around Washington, a vicious storm attacked. It was dusk on a Friday night in June. Winds gusted up to eighty miles per hour. The "derecho" tore into the grid.

Then there was the darkness, the heat, and the silence of the grid undone. By Saturday morning, bad news started to pour in from the region's utilities. Amazon Cloud went dark—Netflix, Pinterest, Instagram too. Washington's subway quit, Amtrak as well. Traffic lights and almost everything else stopped.

Some 1.3 million households and businesses powered up by Washington-area utilities that Friday afternoon powered down with the sunset. Over 1 percent of the country felled in a single attack. More than twenty-four hours later, Saturday night, the utilities estimated 1.1 million customers remained down. More than forty-eight hours later, Sunday night, one utility estimated nearly half its customers were still down.

Into day three, the number of lines down had only been dented. The utilities, overwhelmed with limited resources immediately on hand—linemen, construction personnel, tree cutters, engineers, trucks, tools, spare parts, etc.—naturally prioritized reconnecting the most critical facilities first, like hospitals.

Next, the utilities took on those repairs that would reconnect the greatest number of customers. Makes sense, right? Fix circuit A. One thousand customers would come back up. Fix circuit B. Only one hundred would come back up. So circuit B's repair is skipped over for now in the utility's triage.

The prediction was that it would take a week to reconnect all of the customers. The hundreds of households and businesses on those second-tier circuits and tens and ones on third- and fourth-tier circuits were sorted to last in the triage. But a week? Why should it take that long? Why did it take longer?

Utilities have an admirable tradition of mutual assistance. They help out their neighbor utilities in time of need. When a storm attacks and a utility is overwhelmed repairing its corner of the grid, volunteer crews travel hundreds and thousands of miles to help.

But shouldn't utilities have more resources on hand in the immediate aftermath of an attack? So much dependence on volunteer crews from hundreds and thousands of miles away ensures lines will remain down for some customers for many days, for a week, possibly longer.

It takes time for the crews to drive in. It takes time for them to set up in neighborhoods they don't know well to repair equipment they've never tended. It takes time for the host utility to get everyone organized, to supervise the battalions of volunteers.

Washington pathetic, powerless

Washington, capital of the world's superpower, had been brought to its knees, pathetic, powerless. How could this happen? Seoul withstands vicious storms all the time. Typhoons don't lay flat Seoul's grid for a week.

In the last twenty years, since the Energy Policy Act of 1992, utility regulators have doggedly pursued the goal of driving utility rates down to their absolute minimums. Well, the quest has consequences. We see the result of this rates myopia after each and every storm.

Utilities have responded as anyone could have predicted. They have cut costs, again and again, year in and year out, looking to do more (or as much) with less.

For a while, constant cutting didn't matter much. In the early 1990s there may have been ample areas in utility operations for cost efficiencies.

After a while, after twenty years, constant budget-cutting mattered. There aren't ample areas for cost efficiencies any longer.

The thin reflective orange line of ready utility crews is our forward line of defense against the blackout barbarians assembling at the borders. When a storm attacks, after twenty years of cutting, our forward defense is thinner. It can't hold the strongest attacks. Reinforcements—local contractors and mutual assistance—must be rushed to the rear, belatedly, to where the line has fallen.

It takes longer to restore everyone's power when only a fraction of the crews that are needed are on a utility's payroll and when only a fraction of the equipment that is needed is in a utility's garages and warehouses. Those who have single-mindedly pursued minimal rates can take comfort that the thinning of the reflective orange line is exquisitely economical. Much like the thinning of the blue line of our police forces is economical, until the next crime wave.

Mutual assured destruction

The electric utility industry is proud of the mutual assistance tradition, rightly so. Volunteer crews travel hundreds and thousands of miles to help overwhelmed utilities recover from storms. The dedication and courage of the volunteers are repeated several times every year.

An admirable tradition it is. But the tradition is not one that has been with us for very long. Up until the 1990s, mutual assistance was rare. And it was unheard of for volunteer crews to travel many hundreds of miles, let alone thousands.

Not because utilities in those days were any less supportive of each other. Not because utility crews were any less heroic. Not because the storms were any less ferocious. No, mutual assistance was rare before the last twenty years because it just wasn't as needed.

What's the difference between before and now? Now, mutual assistance is the bastion in utilities' storm defense, as in the aftermath of Superstorm Sandy in 2012. Mutual assistance crews that hail from far-flung utilities across North America outnumbered local utilities' workforces. That wasn't true before, as in the aftermath of Hurricane Andrew in 1992, one of the most ferocious destructive storms ever. Mutual assistance crews that summer from the region's utilities were valuable but limited reinforcements for local utilities' sizable workforces.

Why has mutual assistance grown from reinforcing units that are appreciated to main battle divisions that are crucial? Has this growth led to faster repair of the grid's wounds following storms?

The difference is simple. Utilities now have limited in-house resources. They can no longer combat storms on their own without considerable outside help. Mutual assistance has grown to fill the void.

Conscious choice

Utility regulators have put unprecedented pressure on utilities for twenty years. Regulators now regularly rip into utility proposals like a derecho rips into the grid. They demand that utilities slash costs to minimize customers' rates.

Sure, anyone prefers getting a lower price for anything, electricity included. You'd be a fool not to. But anytime a lower price is on the table, there's always a big if. Getting a lower price is great but only if you get back the same quality and value in the good or service you're buying.

There's a lower price for your favorite breakfast cereal. That's cool. But if the cereal box has been made smaller, that's not so cool. It's worse if the savings from the lower price is but small change.

There's a lower rate for your grid electricity because utility regulators hung tough in the last rate case. That's cool. But if the grid is now more susceptible to lengthy blackouts, that's not so cool. As with cereal, it's worse if the savings from the lower rate is but small change.

Listen to what some regulators have said over the years. It's as if lowering rates were the main object of electric utility regulation. What about what that might do to the grid's quality and value? It's as if lowering rates were a free lunch without a bad aftertaste of more and lengthier blackouts.

At the same time that utility regulators have insisted on minimal rates, federal and state governments have compelled utilities to spend more money on this and that beyond local grid reliability. Keeping rates minimal, utilities are required to spend more to limit the environmental footprint of power plants, to encourage renewable power plants and energy efficiency, and for new taxes and fees. At the same time, utilities spend more on interstate high-voltage transmission lines, smart meters, and personnel benefits.

What happens when you insist on minimal rates while at the same time the cost components of rates, other than for local grid reliability, rise? Where do you think that leaves the cost component for local grid reliability? Uh-oh.

The regulatory process worked, kinda. It forced utilities, after twenty years, to the result regulators sought. Utilities slashed costs anywhere they could. Forget growth in the number of customers. Forget growth in the usage of electricity per customer. Forget added environmental and labor safety regulations. Forget inflation in commodities, equipment, construction, and health and retirement benefits. Forget all that. Utilities would become slashers.

Utility regulators have gotten exactly what they wished for. Utility payrolls have fewer linemen, engineers, and construction workers. Utilities are lean and mean. They are so lean and so mean they have no choice when storm winds pick up. They must call for contractors and mutual assistance to close the yawning gap.

It was a conscious choice made by twenty years of utility regulators. They might have chosen higher utility rates, as a reasonable tradeoff for customers, to equip utilities with sufficient storm restoration capability. In the Southeast, utility regulators may have done that.

But beyond the Southeast, utility regulators generally chose to minimize customers' rates knowing full well utilities compensated by slashing costs and capabilities. Utility regulators knew or should have known that, eventually, diminishing utilities' capabilities would have real consequences for customers. Well, eventually has arrived.

Lean consultants

The Shinkansen, the pride of Japan Rail, sprung out of Nagoya Station, soon to reach a speed of 165 miles per hour. I'd just inspected the gigantic wall being erected to make sure no tsunami, no matter its ferocity, could threaten the Hamaoka nuclear power plant. I looked out my window to some of the works of Toyota. Ah, Toyota. That's where the philosophy of lean became a fundamental principle for industry worldwide. Lean became the way for all kinds of companies to go, hailed as a new salvation by consultants and business writers, eventually holding sway over America's electric utility industry.

Companies adopting the lean philosophy cut costs significantly by cutting their ready on-property resources down to the bare necessities. Lean companies count heavily on outsourcing workforce and just-in-time equipment to make up the difference.

A lean company cuts workforce on its payroll to the bone, counting on last-minute calls to temporary workers and contractors to meet the occasional surges in staff needs. A lean company cuts equipment stored in its warehouses to the bone, counting on last-minute deliveries from vendors to meet the occasional surges in equipment needs.

When a lean company's sales temporarily spike up, resources are spiked up to sate the surge. Resources—workforce and equipment—are afterwards geared down just as quickly. This way there is no slack once sales revert to normal levels.

Responding to regulators' demands to minimize customers' rates, lean naturally seemed to be a welcome panacea to utilities juggling the conflicts of minimizing rates versus escalating costs. Lean offered the proverbial free lunch that utilities could do the same with less.

Instead of maintaining a workforce that was more than ample on good weather days, shrink it to a number that is just sufficient for those days (and not one body more). Instead of maintaining a store of equipment that was more than ample on good weather days, shrink it to a level that is just sufficient for those days (and not a spare part more). A lean utility is sized to days when the lines are up and humming.

Fallacy of the classic inventory problem

To consultants and business writers, the logic in all this comes from the classic inventory problem. Taught at every business school in the country, the classic inventory problem goes something like this.

A company sells a certain number of an item each time period (a day, for example). It takes in shipments of the item periodically. In order for the company to maximize total profits, how should it trade off the profit of making sales from a well-stocked inventory versus not over-ordering shipments lest some excess stock sits around idle for a while? It turns out the mathematics of the situation show there is an optimal quantity of items in each shipment, maximizing total profits.

However, the logic of the classic inventory problem, the underpinning to the lean philosophy, inaptly applies to the challenges facing electric utilities. In the language of the classic inventory problem, the number of an item that electric utilities sell each time period is extraordinarily volatile and unpredictable.

On the vast majority of days, the weather is good and the grid hums. Sales, in the sense of the inventory problem, are moderate. On a minority of days, though, without warning, the weather turns vicious and the grid gets badly wounded. Sales, in the inventory problem sense, dramatically spike up, through the roof.

The challenges facing electric utilities are more aptly defined by another topic covered in the country's business schools. The classic inventory problem has a cousin, the classic umbrella sales paradox. The cousin better represents the challenges facing electric utilities.

If a retailer of umbrellas were to stock up only for the normal fair weather day, it would lose out big-time on the occasional foul weather day when customers stream in, begging to buy cover from the rain. So, instead, the smart retailer overstocks, trading off the costs of overstocking for 365 days a year versus the profits of those occasional days of downpours when demand for umbrellas sharply spikes up.

Now let's consider the umbrella sales paradox applied to the challenges facing electric utilities. Like the overstocking on umbrellas, it might seem wasteful to keep on-hand a more-than-ample number of linemen, engineers and construction workers, and amount of their equipment for 365 days a year that exceeds what are needed on normal days.

Until the occasional blizzard or thunderstorm, all that personnel and inventory and more are sorely needed. On those all-hands days, the more-than-ample resources that are already on property and ready to roll can be decisive in shortening the times that customers are suffering the loss of electrical service.

By the same superficial logic, it might seem wasteful to keep on hand a more-than-ample number of firemen, police, emergency room doctors and nurses, and the armed forces, for that matter, and a more-than-ample amount of their equipment, for 365 days a year. Until... The leaderships of these organizations know full well that, even with the provision for off-property backup and reserves, there is no substitute for already on-property and ready-to-roll resources to combat and contain the occasional emergency.

Fire departments, police departments, hospital emergency rooms, the armed forces, and electric utilities are all applications of the umbrella sales paradox. Keeping extra firemen on the payroll makes perfect sense even though they may sit idle a vast majority of the time. When a major fire does strike, all those firemen playing cards in their little red firehouses, and more, are needed to meet the challenge. Lean would be mean for those the firemen will be rescuing.

Reliability riff

Electric utilities are always talking about reliability. Reliability this, reliability that. It's stated as utilities' top goal ahead of any others. It's said to be a solemn promise to customers. It's job number one.

Okay, then, utilities must really think reliability is important. But what the heck is this thing?

Look it up in the dictionary. Reliability is consistency in quality or performance and trustworthiness. Count on us. We're solid, like the Rock of Gibraltar. We're always there whenever the need arises, like Mom. We're not flashy but relentless, like Steady Eddie.

But an electric utility company? Reliable? The utility's poles don't splinter and come crashing down onto the streets, shearing cars in half. This is true. The utility's substations don't explode, throwing shrapnel of electrical equipment parts all over the neighborhood. That's good. The utility's wires don't spark and sparkle like cheap fireworks as they loop into our houses. Fine enough.

Still, reliability can sound awfully amorphous and abstract when utilities get into their reliability riff. Goliath Power & Light Company claims its dedication to reliability. Uh, what's that supposed to mean?

Plug in the iPad for a fill-up. Yes, this does always work time after time. Nudge the charger plug all the way into the wall outlet, flush with the wall for a nice, snug fit. After a little while, never fails—the iPad battery is topped off.

Though that strikes few as remarkable. It hardly seems worth crowing about, though Goliath P&L does. Should we be that impressed? Should we be all that pleased the wall plug doesn't fail every once in a while?

Everyone knows what's behind a wall outlet. Wires snake around within the walls, all hot with current. Plug a machine, appliance, or device in an outlet, and the juice flows automatically. So? Not real complicated.

Almost exactly like giving a turn to the kitchen sink's cold-water knob. The water flows automatically. So, dear Goliath P&L, it's not rocket science. Reliable, okay. But it's not that big a deal.

This utility talk about reliability sounds a bit too overblown to most people, a bit too lofty, a bit too reverent. Bordering on puffery to some ears.

If there is really something there, something genuine, Goliath P&L and fellow utilities should come right out with it. Utilities should say it plainly, not obscure what this reliability thing is.

If reliability is actually the top goal of the Goliath P&L crowd, if it is actually so important, people naturally feel utilities should lay it out there. Then the public can judge for themselves the worth of what utilities are saying they so diligently pursue.

On the defensive

That utilities aren't more explicit about this reliability thing is confounding at best, aggravating at worst. The coyness and fuzziness as to what reliability exactly is clues many in the public to doubt utilities' consistency and trustworthiness. A reaction on the part of customers that is directly opposite to what utilities desire and intend. Not good.

Is there any there there? Or does all of the reliability talk just mean utilities don't have much of a substantive message?

Running utilities must be so routine, so matter of fact, many in the public must conclude. Same old same old. Watching a bunch of dials day in, day out, checking that all of the juice is flowing around to all the right places. And cashing all of the checks people send in to pay their electric bills. Maybe reliability is the best you can say about it.

Utilities do have to come up with something to say about themselves. People get that. There's no mystery about utilities wanting to boost their image, just like normal businesses.

When utilities demand new rate hikes, the familiar trope about reliability is really rolled out. "Here at Goliath P&L, we're reliable. So when we say a hefty rate hike is necessary, you can trust that the additional money is truly needed. Honest."

You just know Goliath will put the additional money to good uses, say the utility spokesmen. We'll not take a penny more than what is absolutely mandatory to keep Goliath as reliable as ever.

Unpleasant truth

This amorphous talk about reliability is bewildering. Why do utilities rely on the reliability euphemism? If you think about it, their coyness is perfectly understandable. Utilities reasonably shy away from being specific.

You would, too, if you were in their shoes. If you're talking about reliability, you're dangerously close to talking about unreliability. If you're talking about unreliability, you're dangerously close to talking about...

Blackouts. And if you were in their shoes, your utility's shoes, would you want to talk about blackouts? No way! It would be like your surgeon wanting to talk about malpractice suits. Not exactly what your surgeon wants to be doing.

If that dreaded "B" word comes up, people naturally want to know if there'll be blackouts going forward. The answer is, "You bet," but nobody wants to say that.

There'll be blackouts going forward, that you can rely on. And they'll be ever more painful as time goes on. Nobody wants to say that—nobody.

With our ever-increasing dependence on the continuous flow of electricity through the grid's arteries, veins, and capillaries to us, the pain of interruptions will become ever more acute. More pain, less gain. Who wants to come out and say that?

My generation grew up in a semi-electrified world with electrical conveniences like lighting, a television in the living room, a refrigerator in the kitchen, and elevators in tall buildings. Our children's generation is growing up in a hyper-electrified world saturated with electrical equipment all around them. When they reach my age, how will they be able to tolerate blackouts lasting days? It's a terrifying thought.

The unpleasant truth is actually more unpleasant. Not only will blackouts keep happening. Not only will they become more painful.

No point in sugar-coating the situation; it's really a degree worse. Blackouts will likely become more frequent. Blackouts will likely become more widespread. Blackouts will likely become lengthier. All while our dependence on the continuous flow of electricity is ever increasing. Uh, that's bad.

We can do this

If we committed to making blackouts less frequent and less lengthy, could we pull it off? Is it within our abilities? Of course it is.

It's technically feasible to make power outages fewer and shorter, clearly. The technology and process changes needed to make a significant improvement in reliability are well understood and proven. Utility regulators have known this or should have known this.

It's economically feasible as well. The investment needed to make a significant improvement is small relative to the outage costs that customers incur.[54] Utility regulators have known this, too, or should have known this.

Doubling is tripling

Some have assumed that the impact of blackout reduction on customers' rates would be catastrophic. Rates might even double. But that's crazy.

Less than half of a utility's costs, often much less than half, are for maintaining the local grid. More than half of its costs, often much more than half, are for power plants, the regional grid, and government taxes and fees. Understanding this split between local grid costs and other costs is key to understanding the craziness of rates doubling for blackout reduction.

We'll again illustrate with an example. Suppose Zeus Lightning Company spends $2 billion annually for maintaining the local grid. It spends another $2 billion annually for power plants, the regional grid, and government taxes and fees.

So Zeus spends $4 billion annually in total. It collects that same $4 billion from customers paying their monthly utility bills.

54 Kathryn Siegel, "The True Cost of Power Outages," *Yale Environment Review*, Aug. 6, 2012. "Currently, death tolls from disasters only include accidents, like drowning, which can be directly connected to the tragedy. Yet, this approach greatly discounts the severe stress on health that blackouts induce. People are trapped in subways and elevators. Residents in high-rises needed to climb flights of stairs. No air conditioning was to be found and cell phone service was disrupted. Stores and pharmacies closed. Ambulance response time increased and hospitals became overcrowded. Until non-accidental deaths are attributed to power outages, the actual costs of blackouts are not being considered. As governments plan for natural disasters and seek to build more resilient infrastructures, these true costs should not be ignored."

Suppose businesses pay 60 percent of what Zeus collects, $2.4 billion. Households pay the remaining 40 percent, $1.6 billion. Zeus serves 1.6 million households. Each household, on average, therefore pays $1,000 in utility bills annually.

Residential rates, we'll assume, are a dime per kilowatt-hour. Each household, on average, therefore takes 10,000 kilowatt-hours from the grid annually. On a daily basis, households pay an average of $2.74 and take 27.4 kilowatt-hours.

The Greek god Zeus had no master. But the utility Zeus does have a master, its regulator. Suppose its regulator doubles rates, aiming to give Zeus ample monies to reduce blackouts.

Residential customers' rates go from a dime per kilowatt-hour to two dimes. On a daily basis, what households pay goes from $2.74 to $5.48. The same doubling takes place for business customers' rates.

With this doubling of rates by regulators, Zeus now collects $8 billion annually. As if it won the lottery, Zeus has $4 billion more than it had before. It can spend $4 billion more for maintaining the local grid to reduce blackouts.

Here's the point. Zeus was spending just $2 billion on the local grid before this doubling of rates. If it can now spend $4 billion more on the local grid, Zeus has triple the money it had before for this purpose.

No way can Zeus spend that much money on blackout reduction. To do so, Zeus would have to erect gold utility poles.

So a doubling of rates translates into a tripling of the monies that a utility would have for the local grid. Significantly reducing blackouts requires nowhere near that much money.

Value of twenty-seven cents

Some have assumed the impact of blackout reduction on customers' rates would be catastrophic, even if rates increase by just 10 percent. They feel an increase of even this magnitude would be ruinous. Come on.

Suppose Zeus's regulator increases rates by 10 percent to reduce blackouts. Residential customers' rates go from ten cents per kilowatt-hour, that dime again, to eleven cents. On a daily basis, what households pay goes from $2.74 to $3.01. The same 10-percent increase takes place for business customers' rates.

Households now have to put another twenty-seven cents daily into the kitty for blackout reduction. That's not a lot for most households. That's really not a lot for them if what they get back in return is a good value. Getting less frequent and less lengthy blackouts would constitute a good value for many.

With this 10-percent rate increase by regulators, Zeus now collects $4.4 billion annually. Zeus has $0.4 billion more than it had before. It can spend $0.4 billion more for maintaining the local grid to reduce blackouts.

This is a pretty good amount. Zeus had $2 billion before for the local grid. Now it has $2.4 billion for it. That's an increase of 20 percent. It's a lot.

Zeus engineers can implement a mix of proven strategies to make blackouts less frequent and less lengthy. They can increase tree trimming around overhead lines, reinforce wooden utility poles, install metal utility poles, underground selective lines, add line redundancy and switchover capabilities, augment the in-house linemen workforce, locate more replacement equipment locally, etc. The mix will significantly improve local grid survivability during storms by making grid components less susceptible to attack, less vulnerable to damage if attacked, and more recoverable if damaged.

Closer to perfect

Utility regulators, anytime they want, can order the utilities under their rule to make it happen. Make blackouts less likely. For those you don't prevent, make them as brief as possible.

Utilities can't attain perfect power for utility customers. We get that. A world without power interruptions isn't achievable. Fine. But utility customers need and want to be a whole lot closer to perfect power than where we are today. So, utilities, move us closer to perfect.

Utilities in Western Europe and Eastern Asia do this. There's no reason why US utilities couldn't.[55]

55 Jonathan Fahey, Associated Press, Mar. 5, 2013. "Experts say average U.S. reliability will probably never be as good as it is in Japan or parts of Western Europe because the country is so big and its population so spread out. But experts say the frequency of outages can still be cut, and there is ample room to reduce the amount of time customers are without power after outages."

But US utilities aren't likely to be authorized by regulators to make these investments at the scale that would be necessary. No matter how vital the continuous flow of electricity becomes, regulators do not look like they're going there. Why not?

Utility regulators accept that blackouts happen. They accept a certain number of them are just going to occur. And that, when they occur, it will just take some time to get households and businesses back online.

A large proportion of utility customers would undoubtedly find this acceptance of the status quo incredibly vexing. They would surely complain that this kind of attitude in utility regulation poorly serves their wants and needs.

Nonetheless, the consensus in utility regulation is that some level of outages is acceptable in the service of a higher goal. Customer rates are the higher goal—well, making rates lower anyway.

Reliability alphabet soup

Utility regulators and utilities know, but seldom acknowledge, that blackout frequency can be reduced. It really can. How often the grid fails is not an immutable characteristic of nature. The grid is a manmade machine. It can be kept tip-top or topsy-turvy.

SAIFI, the obscure acronym for frequency, can be made lower. Lift grid investment, you lower frequency. Grid engineers can do this.

And regulators and utilities know that blackout duration can be reduced. It really can too. CAIDI, the obscure acronym for duration, can be made lower as well. Again, lift investment, you lower duration. Grid engineers can also do this.

But the prime directive of utility regulation is to minimize customers' rates. It's not to minimize customers' SAIFI and CAIDI. So the regulatory process effectively sets blackout performance at a middling level. It's presumed acceptable so long as we maintain minimal rates. That's what utility customers, all of them, really want. Right?

But it seems presumptuous. Doesn't it to you? Regulators, when deciding to minimize blackouts on the one hand and minimize rates on the other hand, apparently harbor no doubts. They err, of course, on the side of rate reduction.

Blackouts/rates tradeoff

How can utility regulators feel so self-assured they have the tradeoff right? How do they know they've balanced blackout reduction and rate reduction? And how do they know they've set the balance as utility customers would?

Some are happy to hear that regulators err on the side of rate reduction. They want utility rates minimized even after learning blackouts will become more frequent and lengthier in their communities. Call these households and businesses the rate-sensitive customers.

But some aren't happy to hear how regulators err. They want utility blackouts minimized even after learning rates will become higher. Call these households and businesses the outage-sensitive customers.

And some utility customers are in between. They don't want regulators to go overboard on rate reduction, risking reliability. They don't want regulators to go overboard on blackout reduction either, changing course from rate reduction so radically. Call these households and businesses the balance customers.

None of these three broad categories of utility customers is right or wrong. They each have different needs and wants when it comes to the tradeoff between blackout reduction and rate reduction. That's a challenge for utility regulators and utilities. They must conjure up a level of investment in the grid, and an allocation of that investment, that best serves the needs and wants of all three groups.

Reddy's Recharges

1. Utility customers incur significant and exponentially increasing blackout costs as the duration of a power outage increases into multiple days.

2. The statistical distribution of utility customers restored after a major storm is skewed, as in other areas of electricitynomics, with a right-side tail representing multi-day outages.

3. Utility regulators have heretofore neglected the considerable costs to customers of multi-day outages, putting a higher priority on rate reduction than on blackout reduction.

"Get in, Tom—we're goin' to Oklahoma. Hear tell they got 'lectricity."

Chapter 10: Refocusing on Value for Money

Regulators are group buyers; pressures on regulators from those with agendas; utilities must talk up grid quality

Shopping experience

It's an old saying in the industry. A utility has one customer. It might have millions of households and businesses that take grid electricity from it. Sure. But none of those millions makes what we would call a purchase decision. Few know how much electricity they're buying. Almost none know the price.

There is no shopping experience for grid electricity. When you move, your home comes complete with a line to the local power company. All 120 million homes in America have lines already. New homes are automatically electrified. Existing homes have had the same lines to the same utilities for decades. Even if the utility's name changed over the years, it's essentially the same outfit.

Nobody moves to one community or another because of its utility. Once you're situated, nobody chats with their new utility to get acquainted.

One utility powered my parents' home, and a different one powered my dad's store just a few miles away. My dad didn't care. He barely knew.

My dad couldn't have told you which of his two utilities had the higher price. He couldn't have told you whether he took more grid electricity at home or at the store.

You see, we're connected to our utilities, but really we're disconnected from them. Almost all the time, they're far in the background of our everyday lives.

Well, then, who's the real customer? Why, it's the utility's regulators. They're the ones who make what we would call a purchase decision, kind of.

Group buyers

They're more like a buying group. The regulators get into the weeds of quantity, quality, price, and terms and conditions on behalf of the larger group. So in that regard, yes, they're the utility's customer.

Group buyers vary. Some bring more value to the table, to their groups, than others. The very best buyers bargain for the best package—for the best value—in the eyes of the group members.

Others fall short. Some simply don't bargain as well, leaving value on the table. But many lose touch with the needs and wants of the group members. They bargain for a package that isn't the most fitting. They overly consider their own priorities, putting those above the members' priorities.

Congressmen, for example, are group buyers for their constituents back home. We all know that some congressmen lose touch with the needs and wants of their districts as they circulate among cocktail parties and lobbyists currying favors in Washington. They increasingly lean toward what's popular in the capital rather than on Main Street in their districts.

Regulators can lose touch as well. In the last twenty years, more have. During this time utilities have laid low on their view of what customers need and want. They've stayed out of the limelight as much as they can. Into this vacuum, activists advocating this or that cause have entered.

Some lobby regulators to go for the lowest price and to forget about the other parts of the package. Some lobby regulators to go for the lowest quantity above anything else. They're passionate about energy efficiency. Still others lobby regulators to add new terms and conditions, like those greasing the skids for home-brew electricity salesmen. After those are put in, they urge the regulators to add some more.

Strangely, there aren't any activists advocating for premium quality, like making power outages rare and brief. Why is that? Well, a natural advocate for high-quality grid electricity is the utility itself. Like we've said, utilities have shied away from the aggressive tactics of the activists for the lowest price, for the lowest quantity, and for more and more terms and conditions.

So, guess what. Regulators have gone with the flow. Wouldn't you? One moment, a utility's price, minimizing it, is a top priority. The next moment, a utility's quantity, minimizing it, is a top priority. Then, before you catch your breath, adding in terms and conditions is a top priority.

Which leads to the question, is that what the households and businesses served by the utility need and want? And if not, shouldn't that matter to regulators as much as the agendas of the activists?

Haggling with the car salesman

There's another problem I've not yet mentioned that stands squarely in the way of regulators giving equal weight to a utility's quality. The ultimate customers of a utility, the millions of households and businesses, are seemingly divided on their quality needs and wants.

If all of a utility's household and business customers desired premium quality, valuing it above price and other parts of the package, the regulators' call would be so much easier. Or if the millions of customers desired minimum price, knowing that meant mediocre quality and the more than occasional lengthy blackout, the regulators' call would be easy too.

But a utility's customers are divided. Just as when the public is presented with most products and services out there. Some prefer premium quality, accepting higher price. Some prefer lower price, accepting compromised quality. And the remaining consumers are in between. They prefer a measured mix of quality and price.

A utility's customers are confused too. They don't have the information to make an honest choice.

The typical household or business doesn't know there's a tradeoff between a utility's price and quality as there is for products and services generally. Of course they don't know. They don't have a shopping experience for grid electricity. They don't sit down with the utility and have a sales conversation in which price and quality are haggled over.

In the new-car dealership, you haggle with the salesman. "I'm absolutely not going over x dollars a month. How about you throw in a sunroof? And while you're at it, I'd really like that upgraded upholstery."

Each time the salesman gets up and goes to confer with the supervisor, you know what they're up to. They're playing off price and quality. "Let's give him this but hold back that."

And they know what you're up to. You're doing it, too, playing off price and quality. "I'll give up this if I can get that."

The haggling is done. You and he end up shaking hands over a deal. You and he came to an agreement about the tradeoff. You were willing to give a little on price to get a little on quality. He was willing to give a little on quality to get a little on price.

That's buying a new car. But this is being a utility customer, for which there isn't a shopping experience. Who knows if there is a price-versus-quality tradeoff? You don't. You're not in that conversation.

You have a faint idea about the utility's price. It's certainly more tangible than the utility's quality. Look at your monthly utility bill. There are various prices in there among all of the fine print. (Can the font size be any smaller?) There are prices in there, though in incomprehensible formats, like $.002142 per KWH for ECR and $.006004 per KWH for SRR. What the heck is that? Is it too high or too low? Is it a good value for money? Nobody you know knows.

But quality, that's something else again. The utility's quality is totally intangible. It's a complete unknown. What does it even mean to say a utility is high-quality or low-quality? The terms seem meaningless. If you travel around the country, you know grid electricity is the same everywhere, even if the prices the local utilities charge aren't.

You move into a new town. Do you ask how good the utility is? Does the real estate agent bring it up? How about the new neighbors? It's hard to picture them saying, "Yeah, and we've got the greatest utility, you won't believe it. The power never even flickers." No way! Nobody says that.

Barely recall our pain

Until the power does go out. Then, then the talk is all about the utility's quality. People hardly talk about anything else.

If the power is restored quickly, everyone's irked but impressed by the hard work of the utility crews. If the outage lingers for days, even if only for a few neighborhoods, everyone's incensed and insistent on change.

Ask, then, if people want the utility to better harden the grid, to limit the damage from storms, and to repair the damage faster that nonetheless takes place. You might come close to a unanimous response.

Then ask if they'd be willing pay a few pennies more daily so the utility can do it. You won't get a unanimous response. A few expect the unreasonable, an invulnerable grid on the cheap. But most will consider it a good value. A few pennies more daily for the peace of mind that future outages will be less frequent and take less time to fix.

People respond differently if the last outage they lived through was months ago or even further back in time. The proportion favoring a more resilient grid and a better-equipped utility will be less. It's obvious why. The distress, disruption, and discomfort of their last outage have faded with time. Back to the routines of everyday life, we can barely recall our pain.

But there's another reason why people respond differently. The probability that the next outages will be severe—barring grid and utility strengthening—is uncertain. Maybe the next outages will be frequent but maybe not. Maybe the next outages will be long but maybe not. Why not hope for the best? After all, many people in flood-prone zones don't buy flood insurance. Many people in high-crime neighborhoods don't install burglar alarms. It may seem illogical to you, but nevertheless plenty of people go this way.

So, distant in time from the crisis of a severe outage, customers can't get too excited about paying the price, even a small one, to mitigate the next outages. Amidst the crisis of an ongoing or recent outage, the contrast is stark. Customers can clearly see with their own eyes the price-versus-quality tradeoff. The value tradeoff is vivid then.

Low-pricers

What are regulators to do? Really, what can you expect them to do?

Regulators are, if anything, extraordinarily patient. They attentively hear out anyone that has anything to say about the utilities they regulate. There are the full-throated demands of those in the business of advocating low price. Virtually any argument this crowd wants to throw out there for why the price of grid electricity that utilities charge should be as low as humanely possible is respectfully listened to.

The low-price crowd opposes any proposal to better fund the utility. Whether there's a package to catch up with inflation, to install clearly needed equipment, whatever, they're predictably against it.

The low-price crowd seems intent on putting the squeeze on the utility until it squeals. Then they want to turn up the pressure some more.

They're myopically focused on lowering utility rates. It's their shtick. The low-pricers can't or don't want to see whether their demands lessen a utility's ability to deliver power all the time to all. The low-pricers assume away the grid's quality.

Will the households and businesses the low-pricers purport to stand for be better off? Or will the public actually be worse off? How confident are these advocates that the public is as willing as they are to trade in some get on reliability for some give on rates?

Low-quants

Then there are the equally vigorous demands of those in the business of advocating low quantity. Virtually any argument this crowd wants to throw out there for why the quantity of grid electricity that utilities sell should be as low as humanely possible is respectfully listened to by regulators.

The low-quant crowd supports any proposal to better fund programs to lower how much grid electricity people use. Whether there's a package to subsidize these programs by raising rates, or by moving monies from utility investment and operations, or a mix of both, they're predictably for it.

The low-quant crowd is myopically focused on lowering use and utility sales. That's their shtick. Just like the low-pricers, the low-quants can't or don't want to see whether their demands lessen a utility's ability to deliver power all the time to all. The low-quants also assume away the grid's quality.

Will the households and businesses the low-quants purport to stand for be better off? Or will the public actually be worse off? How confident are these advocates that the public is as willing as they are to trade in some get on reliability for some give on energy conservation?

Graph 22:

Even a 10-percent rate increase means fifteen to forty cents daily for most households.

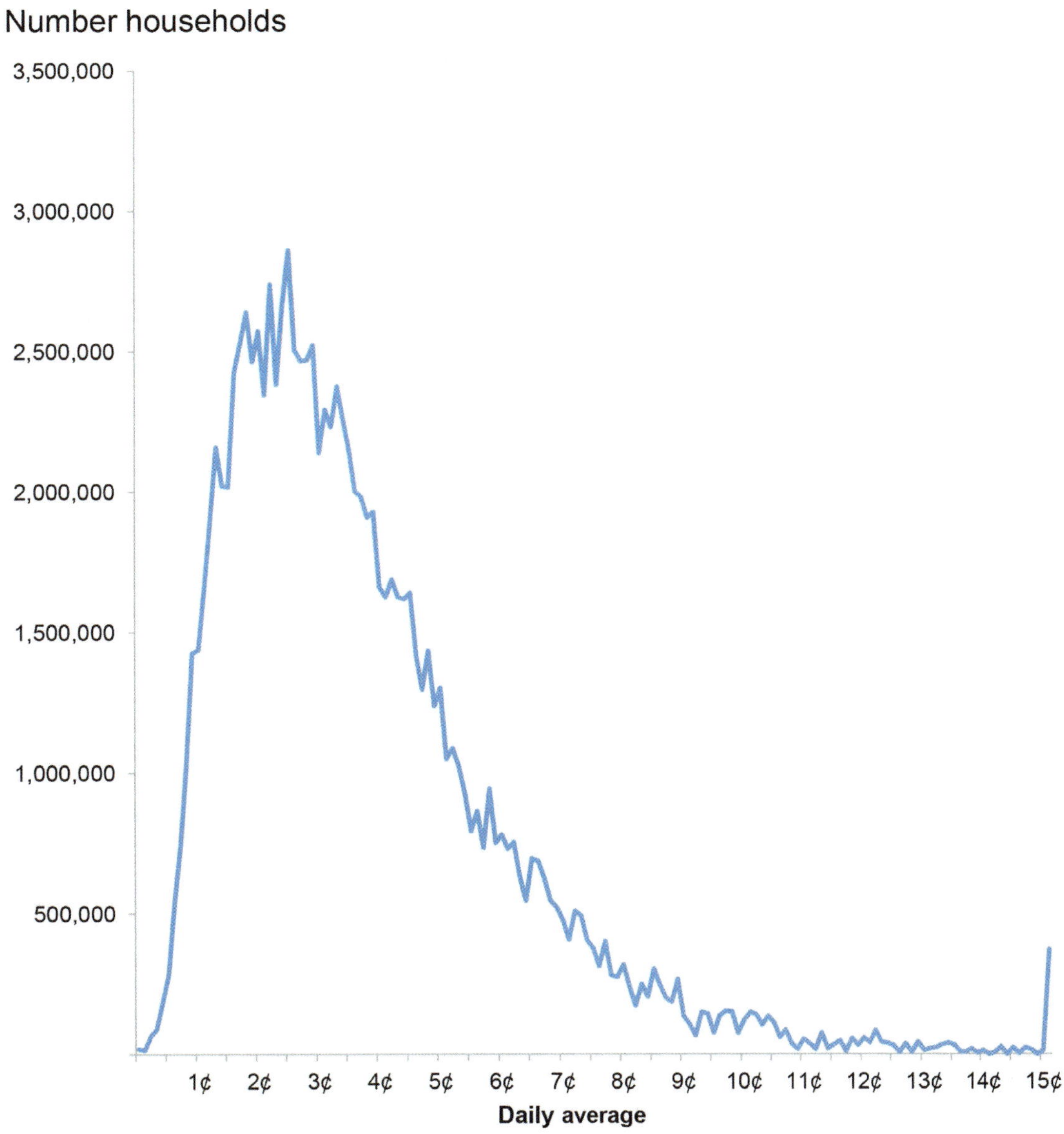

More-T&Cs

And there are also the vigorous demands of those in the business of advocating more terms and conditions (T&Cs) in a state's "regulatory compact" with its utilities. Virtually any argument this crowd wants to throw out there for why there should be more and more constraints and restraints on utilities is respectfully listened to by regulators.

The more-T&C crowd supports any proposal to grease the skids for competitors to the utility and grid electricity, the home-brew electricity installers. Whether there's a package to further tie the hands of utilities from competing with the competitors, or to subsidize the competitors by raising rates, or by moving monies from utility investment and operations, or a mix, they're predictably for it.

The more-T&C crowd is myopically focused on raising home-brew electricity sales. They can't or don't want to see whether their demands lessen a utility's ability to deliver power all the time to all. They may not care. The more-T&Cs also assume away the grid's quality.

High-quals

Who will speak up for grid electricity? Who will stand up for sustaining its quality and raising its quality? Who will step up for a utility's ability—obligation, really—to deliver power all the time to all?

To balance off the low-pricers, the low-quants, and the more-T&Cs, regulators need to hear from the high-quals. If this is not the utility itself, then who is it?

For if the utility won't stand up for what it does for the public, for what it sells to them in the face of increasing competition, what are regulators to do? Regulators naturally will prioritize low price, low quantity, and more T&Cs over high quality. You would too.

And if no one will stand up for the grid, then regulators are fully justified to prioritize quality down to the bottom of the list. Guess what—that's what regulators have been doing for some twenty years now.

Utilities can stand up for grid electricity and its quality on behalf of their customers who so need and want premium quality. But then they must deliver.

It's not enough to urge new monies for more circuits and crews. That won't cut it. It's not enough to make the case effectively demonstrating how customers' value will soar. That won't cut it either. Utilities will then need to go out and deliver on the promise. Only then will their one customer, regulators, and their millions of ultimate customers, households and businesses in their franchise territories, feel they got value for money.

No, utilities must deliver too. They must translate those new monies into tougher grids. Grids that survive more storms—that better happen. Grids that take less damage from the most ferocious storms—that better happen too. Utilities had better deliver.

Discrimination is due

Even industry veterans are surprised when I show them that grid electricity usage and bills and income are strongly correlated.[56, 57] It actually hadn't been recognized that wealthy households take much more from the grid, in general, and pay their utility accordingly. And it hadn't been recognized that poor households take much less from the grid, in general, and pay less accordingly.

Everyone is just as surprised to see that as correlated as grid electricity is with income, it is less correlated than everything else Americans buy, excepting tobacco products. Wealthy households do pay higher utility bills than poor households, but wealthy households pay much more for everything else than do poor households.

This suggests there's some wiggle room on how we spread the grid's cost around among households. I'm no Marxist, but the opportunity is right there for regulators and utilities to spread the grid's cost around so a teensy bit more is picked up by wealthy households and a teensy bit less is picked up by poor households.

Right now, nationally, American households on average spend about 2.9 percent of all of their expenditures on grid electricity. It's not much, so we should pat ourselves on the back

56 The 20 percent of households with the highest income pick up 26 percent of the charges that utilities ask their household customers to pay. The 20 percent of households with the lowest income pick up 14 percent. Source: Bureau of Labor Statistics, Consumer Expenditure Survey.

57 The 24 million households with incomes less than $20,000 picked up just 16 percent of the grid electricity costs of American households, though they're 21 percent of households nationally. The 12 million households with incomes greater than $120,000 picked up 16 percent, though they're just 11 percent of households nationally. Source: Energy Information Administration, Residential Energy Consumption Survey, 2009.

for a job well done. But poor households on average spend about 4.5 percent of all of their expenditures on grid electricity while the wealthy spend just 1.9 percent.

If regulators and utilities were to try a little intra-residential rate redesign, they might manage to drive that 4.5 percent of the poor's expenditures down closer to 4 percent. Indeed, overall funding of grid reliability needs could be increased, at the very same time, by driving that 1.9 percent of the wealthy's expenditures up closer to 2.5 percent. Isn't it just and isn't it reasonable that the wealthy, who take much more from the grid, who arguably depend more on it, and who have a greater ability to pay, pick up a little bit more of the tab?

Inter-class and intra-class

Suppose utilities could persuade regulators that significantly increasing utility budgets to really go after storm-hardening and rapid restoration was good value for money for households and businesses generally. What then? What would regulators do? How could they spread the costs thinly among everyone paying utility bills? How could they spread the costs most thinly among those most sensitive to bill increases? Is it even possible?

It sure is. It's definitely doable to spread the costs thinly and most thinly among the most sensitive to increases. Let's see how.

First off, if utilities and regulators really went after storm-hardening, rapid restoration, and other reliability improvements (including those aimed at briefer power outages), the costs in total will be surprisingly manageable. Utility engineers are generally talented and knowledgeable about grid vulnerabilities. Many already employ value-based planning methods to identify the biggest-bang-for-the-buck projects, the ones that will best patch up the vulnerabilities. My research team has, in cooperation with utility engineers, crunched the voluminous data they keep on circuit outages over the long term. We've seen how you can analyze the data to prioritize the projects that will provide the public the best value for money.

You don't have to underground all of a utility's lines to significantly lessen the incidence of power outages and their duration. Utility engineers know you can selectively underground the most vulnerable and critical lines. You don't have to cut down every tree anywhere near a utility's overhead lines to make a difference. Nor do you have to totally throw out the cost-efficient mutual assistance model and hire thousands more linemen, construction personnel, engineers, etc. Doing a little of everything, using all the arrows in the quiver of grid reliability, can make a big difference.

The costs will also be manageable because the costs for grid investment, maintenance, and operations are only a fraction of anyone's utility bill. The costs for power plant investment, maintenance, and operations, the costs for high-voltage transmission lines, the costs for regional power coordination, the costs for billing and collection, the costs for government fees and taxes, etc., each of these makes up a sizable portion of everyone's utility bills. You could totally go to town, doubling a utility's budget for grid investment, maintenance, and operations, and still only increase the aggregate money paid for grid electricity by around one-quarter. Just consider how much utility engineers could attack the incidence and duration of power outages if you doubled the budgets devoted to that.

Secondly, keep in mind where utility bills are now in relation to grid electricity's value and in absolute terms. Earlier in this book we showed most households pay around $3.00 daily for grid electricity. And many low-income households pay around half that, $1.50 daily. The numbers are lower still outside the Southeast, where households and businesses, too, rely so heavily on air conditioning.

So, even if grid reliability budgets were doubled—which would be a gigantic step and which would increase everyone's utility bills by one-quarter—the cost borne by median households would amount to around seventy-five cents daily. You heard that right. Double reliability budgets. It would literally supercharge utilities' fight against power outages. Yet the cost to median households would be remarkably low, around seventy-five cents daily.

Has the public heard this story? Do they know how much can be done to make outages rarer and briefer for so little cost to them? No, but don't you think they'd like to know?

Double reliability budgets. It would revolutionize the quality and value of grid electricity for everyone. Yet the cost to low-income households, those most sensitive to utility bill increases, would be especially low, around 37.5 cents daily. Don't you think the public would like to know this too?

You don't have to double reliability budgets. Regulators and utilities could go less. Or they could go more if doing so delivered even more value to the public. How should regulators decide how much is enough for reliability or too much? Ah, that's for my next book.

You don't have to double reliability budgets in one fell swoop. Regulators and utilities could phase it in over a few years. That would push down the impact close to the rate of inflation.

You don't have to spread the increased costs for reliability as you have spread the costs of grid electricity previously. Regulators and utilities could slightly redesign rates to soften the impact on low-income households. For instance, rates could reflect the fact that households within densely populated cities are less costly for utilities to serve—in terms of utilities' reliability investment, maintenance, and operations—than are households in sparsely populated rural areas.

It's unusual for a utility to have different residential rates for different communities within its franchise territory. But there isn't any reason why a utility cannot have differentiated rates, if the differentiation is just and reasonable. And if the differentiation isn't unduly discriminatory.

Because it costs less for utilities to serve city dwellers, this would be ample justification to pass the test. In some states where protecting the poor is a high priority, softening the impact on city dwellers—many of whom are poor—might make sense.

Regulators and utilities could slightly redesign rates to soften the impact on industry. For instance, rates could reflect the fact that industry is less costly for utilities to serve—in terms of utilities' reliability investment, maintenance, and operations—than are households and non-industrial businesses. (Industry is typically tied directly into the high-voltage grid, bypassing the low-voltage system that ties in everyone else.) In some states where retaining and attracting industry is a high priority to drive job creation, business formation, and household formation, softening the impact on industry might make sense.

Talk it up

Just once, just one time, let's not hold back. Let's admit and voice the obvious. Grid electricity is a tremendous value.

We can say that. Nobody can contest this simple statement. We might even talk it up.

Grid electricity has become more fundamental to American life than almost any product or service there is. The only things we consume that are in electricity's league are air, food, water, transportation, and shelter. Indeed, as electrification extends into every nook and cranny of our society, those necessities are increasingly based on grid electricity. The air we breathe is the lone exception.

To see this more clearly, picture what happens when the grid doesn't deliver for a time. There's a power outage. Food and water production and distribution is disrupted. Transportation is impeded. Shelter is made inhospitable.

The button I handed out at that industry convention proclaimed, "Electricity is good." This book's introduction opens with the button experiment. But, really, grid electricity isn't just good. It's great.

A contender

As fundamental as grid electricity has become, a contender has now emerged. Government, opinion leaders, and the media have crowned home-brew electricity as an estimable competitor to grid electricity. Making electricity yourself, with solar panels or natural-gas generators, is said to be greener, cheaper, more reliable, and less dependent on Big Brother.

Oh, brother! Can we contain the hype? Some of the claims out there are quite a stretch. They embellish the benefits of home-brew electricity and exaggerate its potential to take grid electricity's place.[58]

Consider, for example, rooftop solar panels. If you have a large house, you might have enough space to put twenty-five panels up there on the roof, each with a capacity to make two hundred watts. Your roof can now manufacture up to five thousand watts of home-brew electricity, up to five kilowatts.

Sounds like a lot, doesn't it? And five kilowatts would be a lot if you could keep the manufacturing process going twenty-four/seven. But, then, that's not going to happen.

During all the nighttime hours, say twelve hours on average, how much home-brew electricity will you be making? Zilch, that's how much. During many of the daytime hours,

58 William Pentland, "Natural Gas: America's Future Electric Grid?" *Forbes*, Nov. 3, 2012. "The interconnected system of transmission and distribution networks commonly known as America's 'electric grid' is broken. 'The North American energy grid is aging, increasingly fragile, buffeted by deregulation and market forces, stressed by relentlessly increasing demand, operating at near capacity with decreasing staffs and reliant on electronic components,' according to Paula Scalingi, an expert on critical infrastructure protection. Today's electric grid was not designed to survive strong winds, storm surges, falling trees and flying debris and seems ludicrously inadequate for the demands of America's increasingly digital and connected economy... Despite spending epic sums of money on the so-called 'smart grid,' the electric power grid seems as stupid as it was before spending billions in federal stimulus dollars. Why throw good money after bad if we have a compelling alternative? And make no mistake about it, we have a compelling alternative to the conventional electric grid. It is commonly called the North American natural gas infrastructure... Of course, the case for using the natural [gas] grid as an alternative to the conventional electric power grid is really just another way of construing the case for distributed power microgrids and all that they entail. DISCLOSURE: I work for a distributed power company."

in the early morning and late afternoon, say six hours on average, how much home-brew electricity will you be making? Only a trickle, that's how much.

During the middle of the day, say six hours on average, if it's a nice, sunny day and if your roof and panels are optimally oriented, then and only then will you make close to that five-kilowatt capacity. You never hear that side of the story.

If you live in Nevada or New Mexico, these optimal conditions will happen often enough. But if you live in Maryland, Massachusetts, or Michigan, good luck with that.

For most of the country, rooftop solar capacity factors—the amount of electricity manufactured as a percentage of capacity over the course of a year—run around 10 to 20 percent.[59] So, if your capacity is five kilowatts and your capacity factor is 10 percent, typical in the northern tier of states, your average electricity output will be half a kilowatt or five hundred watts.

It's not that bad. But, on the other hand, it's only the amount of power your refrigerator operates on. Your main refrigerator, that is. As for powering your second refrigerator, chest freezer, central air conditioning, and everything else in your house, that's still the job of grid electricity.

In your large house, before putting up the solar panels, you might have been taking three kilowatts per hour from the grid, on average, twenty-four/seven. Now that you've put up the solar panels, you've cut your take from the grid by one-sixth, from three kilowatts to 2.5 kilowatts.

In other words, with your solar rooftop, one-sixth of your home's electricity is coming from home-brew, and five-sixths is still coming from the grid. Again, you never hear that side of the story. You only hear the sunny side of the solar roof story.

All the power to them

Knowing the straight facts, some people will want to erect solar roofs anyway. I say all the power to them! It's a free country. But people shouldn't be misled that solar roofs in

59 Renewable Energy Massachusetts LLC, "Massachusetts: A Good Solar Market," www.remenergyco.com/why-solar-now. "The 'capacity factor'(footnote 1) for Massachusetts solar facilities is in the range of 13–15%...This equates to approximately 75% of the solar yield of California [solar photovoltaic] facilities, where capacity factors are 17–20%."

Baltimore or Boston or Brooklyn or Buffalo will take them off the grid. They'll still be taking plenty from it.

And they shouldn't be misled that solar roofs are cheaper or more reliable than the grid. Solar roofs require substantial upfront outlays for homeowners, many thousands of dollars, even after the generous government subsidies.[60]

To lessen the painful expense somewhat, solar roof owners receive regulator-ordered overpayments for any of their excess watts sent back to the grid. Who pays for that? The households and businesses without solar roofs; this subsidy is tucked in their electricity bills.

As for reliability, with your solar roof, will you be laughing through the next power outage? Probably not. When the grid goes down, so, too, does your solar roof, unless you disconnect your house from the grid so as not to electrocute the utility crews and unless you store some power in super-batteries when the sun peeks through the clouds.

Someone should be out there to correct the uninhibited sales puffery from the most overly enthusiastic solar salesmen. Someone should be standing up for the comparative advantages of grid electricity.

Home-brew electricity, the solar way, is greener than grid electricity. You have to grant that to solar roofs—though the actual difference between solar and the grid is slight in terms of their carbon footprint. This is especially true in regions of the country where natural gas power plants are the stalwart of grid electricity. Someone should be saying that too.

What they do

Utilities do grid electricity. That's right. It's what they do—pretty well, as a matter of fact.

A few utilities find other businesses alluring. They long to conduct business without regulators looking over the shoulders. They think to themselves, *Wouldn't that be sweet?*

60 A November 5, 2012, mass mailing by Solar Solution of Washington, DC, stated: "Federal and State incentives for DC residents can cover up to 90% of your solar system. Call the only local solar installer in DC to get started!" The mailing included a table that showed that for a medium-sized 4.4-kilowatt system, governments would subsidize $18,528 out of the total $21,560 cost. The out-of-pocket upfront cost for the homeowner would be $3,032. Even with this 86-percent subsidy, it would still take years for the upfront cost to pay off for the homeowner.

Some utilities find home-brew electricity alluring. Seems like it's the wave of the future. And it seems like it's a similar business. Home-brew electricity can't be all that different than grid electricity.

It is. Success in home-brew electricity requires different competencies than is required for success in grid electricity. The home-brew biz requires the aggressive marketing skills of an insurgency. The grid biz requires the evidence-based regulatory skills of a community institution.

The home-brew biz involves a transactional approach. The grid biz involves an ongoing responsibility approach. Home-brew sellers say to their customers, "When you buy it, you own any of the problems." Grid utilities say, "We own any of the problems."

But that's not the worst part of the home-brew biz for utilities. Does a utility really want to hawk a service that's fundamentally at odds with its core function? Does a utility really want to join the chorus of grid critics further confusing the public about grid electricity's unique strengths?

As the twentieth century came to a close and as the twenty-first opened, the utility industry has gone through several bouts of self-doubt. Who are we? What do we really do for the people we serve?

Some said utilities could sell different kinds of electrons. Some electrons would be greener than others. Some other electrons would have other premium attributes that customers would buy into. But that idea flopped. Consumers didn't get it. Electricity isn't about different electrons in some abstract sense. Electricity is electricity.

Some said utilities could sell different services that derive from grid electricity, like comfort, convenience, and cost control. But that idea flopped too. Consumers didn't get that either. Electricity isn't about derivative services in some abstract sense. Electricity is electricity.

Utilities can, however, sell grid electricity on its own terms. Grid electricity is intrinsically unnoticeable, as this book discussed in its opening pages. This makes it great, and this makes it greater than home-brew electricity. But on those rare occasions when grid electricity becomes noticeable, when the grid goes down, households and businesses suffer severely.

The utility that limits power-outage frequency and duration better than neighboring utilities do would be prized by its customers. Even if the utility's rates were a little bit higher. That is, if the utility's customers knew how high the quality is of their grid electricity.

This is the differentiation that counts with real people. This is tangible and not abstract. This is what utilities do. It's their biz: being such a good custodian of the grid that households and businesses never have to think about any of this stuff.

Age of Electricity continues

Regulators prioritized building out the grid in those first years, the early and mid-twentieth century. The emphasis on building was unquestioned. After all, it was the states' overarching objective when they initiated the franchising and regulating of utilities.

But the country changed by the late twentieth century and so, too, the national consensus about building out the grid. The grid's excesses—particularly air pollution from fossil fuel power plants and then cost overruns from nuclear plant construction projects—were highlighted by critics. Americans soured on the grid.

Determined, able critics demonstrated fossil fuel plants were a major contributor to air pollution. Then they rolled out compelling cases that nuclear plants were uneconomical. Suddenly, the fountain of grid electricity, its fleet of thousands of fossil fuel plants and a hundred nuclear plants, became its failure to many of our countrymen.

Regulators naturally responded to the turnabout. They switched the emphasis in utility regulation. They prioritized reining in the grid's excesses. As the last century turned to this one, regulators sought to slow down the building out of the grid. Some urged regulators to go further and stop the building altogether. That happened to an extent in a number of states.

This retrenching has now been going strong for twenty years. Those grid's excesses, first exposed in the 1970s and 1980s, have been eliminated. Air pollution from fossil fuel plants has been dramatically slashed. Nuclear plant construction was stopped cold. Just as regulators presided effectively during the era of building, they presided effectively during the era of retrenchment.

That was then; this is now. Retrenchment succeeded. There is no rational worry that fossil fuel plant pollution or nuclear plant project overruns will plague the public. Yes, there are real concerns about a more recently revealed pollutant, carbon dioxide (a climate change precursor), but the grid's output of it is falling rapidly.

Are we then on the verge of a new era? Critics and competitors insist the answer is a flat no. The Age of Electricity is in its sunset period.

But that's their opinion. For readers of this book, clearly it's not mine nor must it be yours. You can decide for yourself.

Regulators are the referees. If utilities and you and I stand up for grid electricity, regulators will decide. These good public servants can determine what comes next. They can and should prioritize perfecting the grid's value. This would be a great mission for regulators, in collaboration with utilities, and you and me for the next era of the Age of Electricity.

Reddy's Recharges

1. Grid electricity is a tremendous value for American households and businesses, but if it is not talked up while critics and competitors talk it down, Americans will become increasingly disenchanted with the grid and increasingly unwilling to devote the necessary funds to perfect its value or even to keep its value from degrading.

2. Utilities talked up building out the grid in the mid-twentieth century and then went into a defensive crouch as the grid's excesses were highlighted as the twentieth century turned into the twenty-first but now can and should talk up perfecting the grid's value as we head into the mid-twenty-first century.

3. Regulators prioritized building out the grid in those first years and then prioritized reining in the grid's excesses as the last century turned but now can and should prioritize perfecting the grid's value for the next era of the Age of Electricity.

"Yup. I'm guessing there's your problem."

Epilogue: Grid's Best Days Ahead

Imagine the value

Try to imagine what grid electricity will do for Americans in coming decades. How will an elder's body be augmented? How will a child's learning be accelerated? How will twins stay close while separated by hundreds of miles? How will music, medicine, and manufacturing be transformed? How about public safety and national security?

The coming bounty of future innovations will further exploit grid electricity's unique qualities.[61] It might even become more ubiquitous, inexpensive, invisible, and unlimited in the future. Consider how stimulating that would be for America.

Just imagine if electricity becomes so ubiquitous that you could get some anywhere. Imagine if electricity becomes so inexpensive that it is even more often given away for free.

Imagine if electricity becomes so invisible that the grid is seldom thought about. Imagine if electricity becomes so unlimited that blackouts are rare and brief.

What's in it for utilities?

The progress toward perfect power will eventually make grid outages as rare and brief as water/sewer, natural gas, and telephone/mobile data outages. But what's in it for utilities?

---

61 Brian X. Chen, "New Breed of Robotics Aims to Help People Walk Again," *New York Times*, Sept. 11, 2012. "Ekso is one of several companies and research labs that are working on wearable robots made to help disabled people or to make the human body superhuman. In 2010, Raytheon released a suit for soldiers that is designed to reduce injuries from heavy lifting. And in Israel, a company called Argo Medical Technologies also makes a robotic suit to help paraplegics walk again. Ekso says it is the first company to introduce a self-contained robotic suit, without any tethers to, say, a power supply. And though its suits for the disabled are now used only in rehabilitation centers, it is looking ahead to a day when they will let people take to the sidewalks, the shopping malls—and maybe even the woods."

For those utilities where household formation will be strong, their regulators will support a healthy pace of investment in both the high-voltage grid—power plants and transmission—to manufacture kilowatt-hours cheaply and cleanly and the low-voltage grid to lessen the havoc storms wreak. What about the others? For those utilities where household formation will be weak, their regulators will still support investment in the low-voltage grid to lessen storms' havoc.

So whether they're in growing regions of the country or not, utilities face a future of great promise. They'll be able to significantly increase value for their customers—households and businesses—under the proven regulatory model refocused on customer value. Utilities that do this well will significantly increase their shareholders' value too.

About the Author

Fancy titles

I was heartbroken. At the same time, I felt an irresistible urge to help the city of my boyhood recover from the wounds of 9/11. So I founded a company with the backing of three utilities. We would install underground high-voltage lines to securely pipe into Manhattan and Queens electricity from upstate.

"Empire Connection" was strongly supported by so many. Environmentalists, city politicians, and labor unions rooted for us, even the *New York Times* editorial board. The project got tantalizingly close to breaking ground. But by late 2004, cold reality could no longer be ignored. No herculean effort would overcome the odds and roadblocks protecting political and financial self-interest.

I soon joined the iconic management-consulting firm McKinsey & Co. Had a fancy title: senior advisor. It was great making a difference for top energy companies.

Should I have spurned the next offer that came? It's easy to say now. But I left McKinsey to become Eliot Spitzer's chief energy advisor. The charismatic new governor committed to my job definition. I was to help get as much energy infrastructure built in New York as possible.

Eight months later I left Albany. I left with far too little of that set in motion and the sad wisdom that New York politics is a blood sport suited best for the slyest and meanest.

So back to management consulting. I joined giant Marsh & McLennan, its subsidiary Oliver Wyman. Another fancy title: partner.

After a while, my mission to help get energy infrastructure built called again. That's how I came to found Build Energy America, and that's how I came to write this book.

At Build Energy America, I've been able to concentrate on rigorous cutting-edge research about the barriers to modernizing and expanding our energy infrastructure. I've been fortunate to make use of the extraordinary research capacities of Bates White Economic Consulting. And, yes, they gave me another fancy title: senior energy advisor.

Schooled

I wasn't born a leading advocate of America tending to its energy infrastructure needs. As a kid, I majored in electrical engineering at the four-year Brooklyn Technical High School. My next four years were at Rensselaer Polytechnic Institute. I continued studying electrical engineering. But ultimately my degree was in physics, doubling up with a history/political science degree. Without pause, I went straight to the Wharton School of Business at U Penn for an MBA.

Before long, I was at Georgetown University, going for a PhD in economics, simultaneous with teaching classes there in economics and statistics as a faculty member, simultaneous with heading an energy consulting firm. Is it surprising I struggled to find sufficient time to complete my doctoral dissertation?

The thesis got longer and longer as I attempted to econometrically explain capital investment in the oil and gas industry. I had passed the PhD orals with an emphasis in econometrics and "public finance," the study of government's role in the economy. Passing the orals proved to be the easy part. Finding the time to polish off the thesis was impossible.

In the late 1980s and throughout the 1990s, I was primarily at the large energy consulting firms PHB or Hagler Bailly, which later bought PHB. I co-headed the main PHB Hagler Bailly division for advising energy companies.

Decades of energy

Over the years, I have published more than twenty articles in energy journals and spoken before more than one hundred energy conferences. The *New York Times*, *Wall Street Journal*, *Financial Times*, Reuters, etc. have often quoted me in their energy stories. The utility regulatory commissions of many states/provinces and the Federal Energy Regulatory Commission have benefited from or at least survived my expert testimonies.

I created one of the principal computer programs for utilities to comply with the Clean Air Act Amendments of 1990, *75 Check*. I created two other popular computer programs for utilities, *Real Time* and *Ramp Up*, to formulate strategies for their power plants.

I have had the great pleasure of knowing and advising many of the top leaders of America's energy industries of the past quarter-century. It is rewarding to know I have made a lasting positive impact on a number of the most successful companies.

Glossary

Block-rate structure: Can be inclining, in which the more grid electricity a utility customer takes, the greater is the per-kilowatt-hour price, or it can be declining, in which the more a customer takes, the lesser is the price

CAIDI: Customer Average Interruption Duration Index, a common measure of a utility's reliability, calculated by dividing the sum of all customers' power outage durations by the number of customers' power outages

Cost allocation: Utility regulators' division of a utility's budget between broad classes of customers (such as residential, small commercial, medium commercial, and industrial) to be covered by them based upon regulatory principles

Electricitynomics: The new economics of electricity usage, cost, and value incorporating the principles of skewed distributions, utility customer variation, income correlations, household formation's role in electricity sales growth, daily cost comparisons, blackout cost as value lost, benefit-cost ratios of blackout reduction, etc.

Grid electricity: Electric utility service to every household and business in America

Home-brew electricity: Solar roofs, natural gas and oil generators, and the like that supplement grid electricity at a small but growing number of households and businesses

Household formation: Critical to a utility's growth of grid electricity sales, the relative proportion of new homes in the utility's area

Kilowatt-hour (KWH): Enough grid electricity to run most American households for around an hour (except very large or small homes and except during the summertime, when air conditioning is used intensively)

Inflation: Prices of nearly all goods and services continually increase over the years, so to properly evaluate the price increase of any one good or service we adjust it with the generalized Consumer Price Index

Mean (average): The sum of all of the numbers in a statistical distribution divided by the amount of numbers; the most common measure of a distribution's central tendency

Median: The number in a statistical distribution that splits it into halves of equal amounts, with half of the distribution having a greater number and the other half having a lesser number; the most reliable measure of a distribution's central tendency

Microdata: The grid electricity usage and costs and demographic characteristics of the individual households surveyed by the Energy Department's Residential Energy Consumption Survey and Labor Department's Consumer Expenditure Survey

Mode: The most common number in a statistical distribution; a measure of a distribution's central tendency

Mutual assistance: A pact among utilities in which one utility will help a neighboring utility when a storm hits the neighbor's grid hard, the practice that has recently been overwhelmed as more widespread storms have hit an entire region's grid hard

Normal: A balanced statistical distribution in which the average is at the distribution's median, with half the distribution members having a greater number and half having a lesser number

Quintile: A one-fifth slice of a statistical distribution where the numbers of the distribution are ordered from the least to the greatest, so there's a bottom quintile, next-to-bottom quintile, middle quintile, next-to-top quintile, and top quintile

Rate design: Utility regulators' setting of specific per-customer, per-kilowatt-hour, per-maximum-kilowatt rates for broad classes of customers (such as residential, small commercial, medium commercial, and industrial)

Rate of return on equity (ROE): A utility's allowed profit rate set by utility regulators, upon which utility rates are set

SAIFI: System Average Interruption Frequency Index, a common measure of a utility's reliability, calculated by dividing the number of customers' power outages by the total number of customers

Skew: An imbalanced statistical distribution in which the average is either greater than or less than the distribution's median, with more than half the distribution's members having a greater or lesser number

Tail: One of the two ends of a statistical distribution, the one with the least numbers in the distribution or the one with the greatest numbers

Undue discrimination: Utility rate policies that utility regulators aren't permitted to implement. Any policies that are more favorable for some utility customers over others must have a just and reasonable basis.

Utility: The only company that is permitted by a state to deliver grid electricity within a specified area

Utility bill costs: A utility customer's payment for grid electricity through its monthly utility bills

Utility blackout costs: A utility customer's indirect costs—out-of-pocket losses and other perceived losses in value—incurred because of grid electricity's power outages

Utility regulator: A state government official on a permanent commission that administers the state's public utility laws

Value: A consumer's personal assessment of the benefit he or she derives from a product or service, absolutely or net of its price and its indirect costs

It's not enough to urge new monies for more circuits and crews. Utilities will then need to go out and deliver on the promise. They must translate those new monies into tougher grids.

Research Sources

AARP Illinois. "AARP, Groups, Urge Lawmakers to Stop Utility Rate Hike Express." News release, May 4, 2011, *available at* www.aarp.org/politics-society/government-elections.

AARP New York. "AARP Stops Electric Price Increase." News release, Aug. 30, 2011, *available at* www.aarp.org/politics-society/advocacy.

Baker, Stephen. "Can Social Media Sell Soap?" *New York Times*, Jan. 6, 2013.

Bonbright, James C., Albert L. Danielsen, and David R. Kamerschen. *Principles of Public Utility Rates*, 2nd ed. Reston, VA: Public Utility Reports, Inc., 1988.

Borenstein, Seth, and Jonathan Fahey. "Officials: CO2 Emissions in US Drop Because of Plentiful Natural Gas." *Scranton Times Tribune*, Aug. 16, 2012.

Cardwell, Diane, Matthew L. Wald, and Christopher Drew. "Upgrade or Clean Up? Hurricane Sandy Alters the Cost-Benefit Calculus for Utilities." *New York Times*, Dec. 29, 2012.

Cauchon, Dennis. "Electric Bills Soar $300 in 5 Years." *USA Today*, Dec. 13, 2011.

Chen, Brian X. "New Breed of Robotics Aims to Help People Walk Again." *New York Times*, Sept. 11, 2012.

Consumer Electronics Association. *U.S. Consumer Electronics Sales and Forecasts 2008–2013*, Jan. 2013.

Costello, Ken. "Should Public Utilities Compensate Customers for Service Interruptions?" Report No. 12–08, National Regulatory Research Institute, July 2012.

Cox, Stan. *Losing Our Cool*. New York: The New Press, 2010.

Davidson, Paul. "Price Jolt: Electricity Bills Going Up, Up, Up." *USA Today*, June 20, 2008.

Diana, Carla. "Talking, Walking Objects." *New York Times*, Jan. 27, 2013.

Edison Electric Institute. *Typical Bills and Average Rates Report* (various semi-annual reports).

Eilperin, Juliet. "Study: Proliferation of Home Air Conditioners Saves Lives." *Washington Post*, Dec. 23, 2012.

Fahey, Jonathan. "Electricity Prices Rise Despite Cheaper Costs for Utility Companies." *Huffington Post*, July 11, 2012.

Fahey, Jonathan. "US Power Grid Costs Rise, But Service Slips." Associated Press, Mar. 5, 2013.

Fox-Penner, Peter. *Smart Power*. Washington, DC: Island Press, 2010.

Freeberg, Ernest. *The Age of Edison*. New York: Penguin Press, 2013.

Gordon, Robert J. "Is U.S. Economic Growth Over? Faltering Innovation Confronts the Six Headwinds." National Bureau of Economic Research Working Paper No. 18315, Aug. 2012.

Hirsh, Richard F. *Technology and Transformation in the American Electric Utility Industry*. Cambridge University Press, 1989.

Kahn, Alfred E. *The Economics of Regulation: Principles and Institutions*. Hoboken, NJ: John Wiley & Sons, Inc., 1971.

Koerth-Baker, Maggie. *Before the Lights Go Out*. Hoboken, NJ: John Wiley & Sons, Inc., 2012.

Kotkin, Joel. "America's Growth Corridors: The Key to National Revival." Manhattan Institute for Policy Research Civic Report, No. 75, Feb. 2013.

Lovins, Amory B. "Energy Strategy: The Road Not Taken?" *Foreign Affairs* 6, no. 20 (1976).

Makansi, Jason. *Lights Out*. Hoboken, NJ: John Wiley & Sons, Inc., 2007.

McGeehan, Patrick. "Con Ed Seeks Rate Increase." *New York Times*, Jan. 25, 2013.

Mitnick, Steven. "Real Value." *Electric Perspectives* (Mar./Apr. 2012).

Morley, Janine, and Mike Hazas. "The Significance of Difference: Understanding Variation in Household Energy Consumption." ECEEE 2011 Summer Study, Panel 8: Dynamics of Consumption.

Mufson, Steven. "Cheaper Natural Gas Doesn't Guarantee Lower Fuel Bills for Homes." *Washington Post*, Dec. 15, 2012.

National Research Council. *Terrorism and the Electric Power Delivery System*. Washington, DC: The National Academies Press, 2012.

Office of Governor Martin O'Malley. "Weathering the Storm." Report of the Grid Resiliency Task Force, Sept. 24, 2012.

Owen, David. *The Conundrum*. New York: Riverhead Books, 2011.

Renewable Energy Massachusetts LLC. "Massachusetts: A Good Solar Market," www.remenergyco.com/why-solar-now.

Sanderson, Bill. "New York Electric Bills to Soar 12%." *New York Post*, Apr. 1, 2011.

Savage, Sam L. *The Flaw of Averages*. Hoboken, NJ: John Wiley & Sons, Inc., 2009.

Schewe, Phillip F. *The Grid*. Washington, DC: Joseph Henry Press, 2007.

Siegel, Kathryn. "The True Cost of Power Outages." *Yale Environment Review*, Aug. 6, 2012, *available at* http://environment.yale.edu/yer/article/the-true-cost-of-power-outages.

Smith, Rebecca. "U.S. Electricity Use on the Wane." *Wall Street Journal*, Jan. 2, 2013.

Smith, Rebecca. "Utilities Revisit Storm Plan." *Wall Street Journal*, Dec. 16, 2012.

Smithsonian Institution, National Museum of American History, Archives Center, www.americanhistory.si.edu/archives.

Solar Solution. Mass mailing, Nov. 5, 2012, Washington, DC.

Tomain, Joseph P. "'Steel in the Ground': Greening the Grid with the iUtility." *Environmental Law* 39, no. 4 (2009): 931–76.

US Bureau of Labor Statistics, Department of Labor, www.bls.gov/data.

US Census Bureau, Department of Commerce, www.census.gov.

US Energy Information Administration, Department of Energy, www.eia.gov/electricity and www.eia.gov/consumption/residential.

US Office of Electricity Delivery & Electric Reliability, Department of Energy, Emergency Situation Reports, www.oe.netl.doe.gov/emergency_sit_rpt.

Vo, Lam Thuy. "American Houses Are Still Much, Much Bigger Than They Used to Be." Planet Money Blog, Aug. 23, 2012.

WABC Radio 77. "Famous Musicradio WABC Stories," www.musicradio77.com/stories.html.

Wotapka, Dawn. "Big Homes Are Back in Business." *Wall Street Journal*, June 7, 2012.

Woodward, Calvin. "Debate Fact Check: How Obama, Romney Statements Hold Up." Oct. 3, 2012, *Huffington Post*.

Images, Photos, and Illustrations

The Reddy Kilowatt images are used courtesy of Reddy Kilowatt Corporation, a wholly owned subsidiary of Xcel Energy, Inc.

The *New Yorker* images are licensed by The Cartoon Bank, Condé Nast.

Additional photos and illustrations are licensed by Dreamstime.

Subject Index

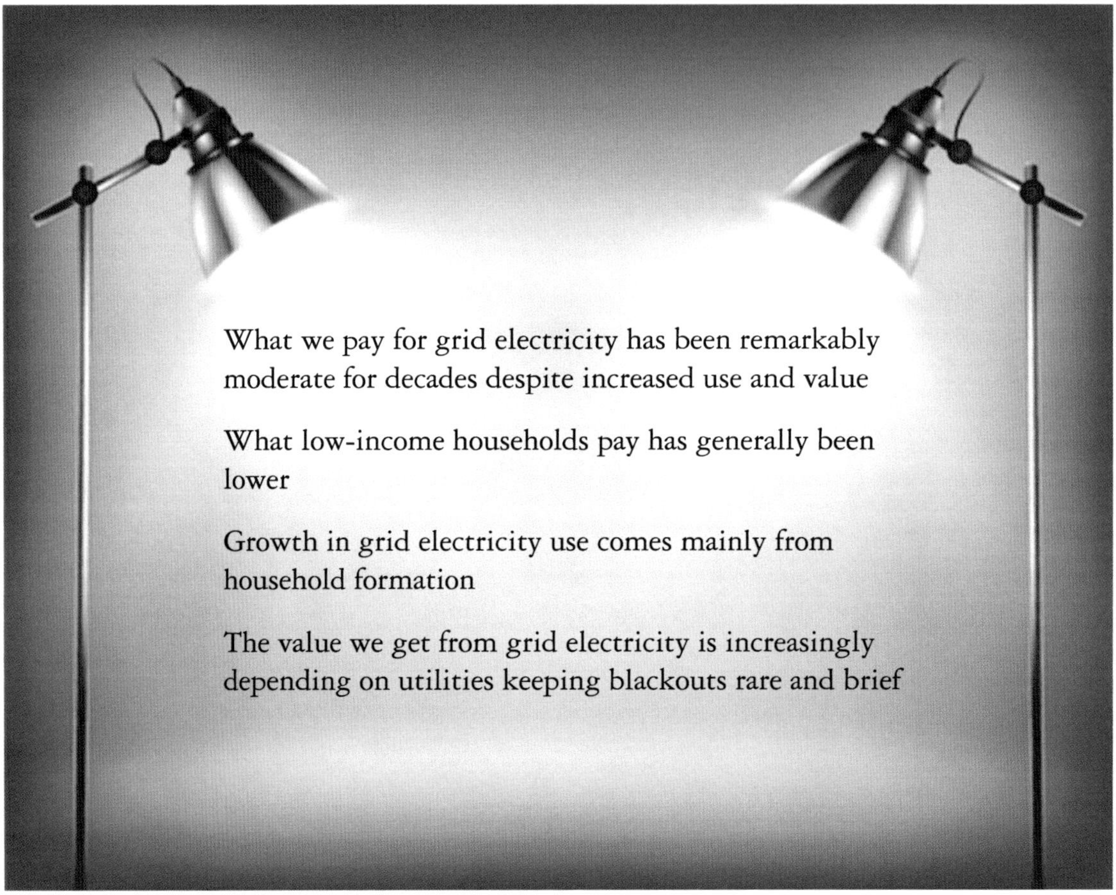
What we pay for grid electricity has been remarkably moderate for decades despite increased use and value
What low-income households pay has generally been lower
Growth in grid electricity use comes mainly from household formation
The value we get from grid electricity is increasingly depending on utilities keeping blackouts rare and brief

27375686R00167

Made in the USA
Lexington, KY